Progress in IS

More information about this series at http://www.springer.com/series/10440

Ahmed Bounfour

Digital Futures, Digital Transformation

From Lean Production to Acceluction

Ahmed Bounfour
European Chair on Intellectual Capital
University Paris-Sud
Sceaux
France

ISSN 2196-8705   ISSN 2196-8713   (electronic)
Progress in IS
ISBN 978-3-319-23278-2   ISBN 978-3-319-23279-9   (eBook)
DOI 10.1007/978-3-319-23279-9

Library of Congress Control Number: 2015947796

Springer Cham Heidelberg New York Dordrecht London

Printed on acid-free paper

Springer International Publishing AG Switzerland is part of Springer Science+Business Media
(www.springer.com)

Foreword

The ISD Program: An Example of Collective Intelligence in the Digital World

The CIGREF Foundation, whose mission is to *better understand how the digital world is transforming the way we live and do business*, rolled out the *Information Systems Dynamics (ISD)* international research program in 2010.

At CIGREF, we believe that digital culture is characterized by sharing of information and knowledge between the different stakeholders of an organization so as to build a collective intelligence that acts as a source of value creation for the enterprise. The ISD program is a remarkable example of the construction of a collective intelligence. When we set the project up in 2010, our aim was to create a collaborative, open program, integrating multiple contributions. In other words, a "research program 2.0".

The program's goals are ambitious:

- Draw on academic research to provide key insights for understanding a different future in a different world;
- Advise the leaders of major public and private organizations on strategic digital issues, in the light of changes under way in business models and in society at large;
- Define new, explanatory theoretical models that offer innovative approaches to IT and digital management, while also helping managers deal with new risks;
- Organize and promote fruitful dialogue between practitioners and researchers, and between business and academia.

The intention, ultimately, is to understand the digital transformations going on around us, to identify the conceptual building bricks of the 2020 enterprise—characterizing its digital uses, its value creation spaces, and how they are governed—and to understand the paradigm shifts inherent in the modes of production of the digital age.

Thanks to the unflagging and enthusiastic commitment of the scientific community around Professor Bounfour, Bruno Ménard and the members of the Steering

Committee, and Alain Pouyat, the chairman of the Strategic Orientation Committee, who met regularly to discuss new ideas and challenges for the program, and thanks to the support of our sponsor companies (Capgemini—Microsoft—Orange—Société Générale—Altran), we have met the challenge of *better understanding how the digital world is transforming the way we live and do business.*

The results of the CIGREF Foundation and of the ISD program have kept the promise of helping us to "find our way" through this complex digital world by providing us with navigation charts to explore possible routes ahead for business in 2020.

Supporting this program is about more than simply creating the conditions for top-flight research. It is about providing all the right conditions to bring the research findings forward into the international arena, making them available to our businesses and our wider ecosystem, including academia and government institutions.

In the 30 ISD program projects conducted between 2011 and 2014—by some fifty research laboratories in the USA, Europe, China and Japan—many topics relating to the 2020 digital enterprise were addressed in great depth. An understanding of these topics is essential to managing our organizations more effectively: business models and ecosystems, mobility, internal innovation practices and open innovation, knowledge flows, data, ethics, norms and standards, economic performance and organizational design.

All of these themes are covered in Prof. Bounfour's monograph *Digital Futures, Digital Transformation: From Lean Production to Acceluction,* a work that also outlines the organizational design scenarios for the 2020 enterprise and presents the concept of Acceluction, a new system of production characterized both by the expansion of the field of value creation to multiples spaces, and by the acceleration of the associated links.

These elements will help us, as managers, to take a clear, informed view of the impact that these issues and challenges will have on our businesses between now and 2020. They will help our organizations to better navigate the new digital world by reconciling economic performance with organizational coherence, by learning to harmonize agility, innovation and collective efficiency, and by mobilizing the values of commitment, cooperation and trust.

Pascal Buffard
President of Cigref, President of Axa Technology Services

Acknowledgments

Writing a book based on the results of an international research program with different disciplinary, national, and functional perspectives is always a challenge. In this case, the task for me as scientific leader and general rapporteur of the program was made easy by the uniqueness of the exercise, as well as by the continuous intellectual and material support of the program's governance bodies.

This is a unique program, where executives and scholars meet together to discuss a scientific and business object: the design of the 2020 enterprise.

As always, the success of a collective endeavor depends upon the willingness and support of various bodies and individuals. I would like to warmly thank CIGREF for initiating and supporting the program. The CIGREF Foundation's Board provided guidance and support, particularly in the early stages of the program when no results were available. I would like to express my deep gratitude to Pascal Buffard, President of CIGREF and President of Axa Technologies Services, who continuously supported the programme's agenda in terms of content, deliverables, and dissemination; Bruno Ménard, Vice-President of CIGREF, and CIO at Sanofi, who greatly ensured the consistency of the dialogue between the programme's deliverables and its sponsors' expectations, and also ensured the dissemination of its main results; to Bruno Brocheton, Vice-President of CIGREF, and CIO at Euro Disney Group, Bernard Duverneuil, Vice-President of CIGREF, and CIO of the ESSILOR Group, Georges Epinette, Vice-President of CIGREF, and CIO at the MOUSQUETAIRES Group, Jean-Marc Lagoutte, Vice-President of CIGREF, and CIO at the DANONE Group, Pierre Laffitte, Honorary Senator, and President of the Foundation Sophia Antipolis, and Alain Pouyat, Executive Vice-president IT and New Technologies, BOUYGUES Group, who, as members of the Board, constantly gave feedback and encouragements to this collective effort.

I wish especially to warmly thank Jean-François Pépin, the *Délégué Général* of CIGREF who worked tirelessly to bridge the research and executive agendas, and who continuously supported and facilitated the program's implementation in terms of its objectives, resources allocation, and dissemination.

Warm thanks also go to the members of the Strategic Committee, who discussed the agenda and the interim results in various arenas and formats, including ad-hoc workshops: to his president, Alain Pouyat, Bouygues Group, as well as to Cyril François, Cap Gemini, Bernard Ourghanlian, Microsoft, Pierre-Louis Biaggi, Orange, Françoise Mercadal Delasalles, Société Générale, and Corinne Jouanny, Altran.

Naturally, this program would have not been possible without the committed support of its Scientific Committee, representing different disciplines (informatics, management information systems, geography, management science, business history, innovation policy, etc.), cultures and locations (Europe, the United States, Brazil, India, China, and Japan). Thank you to those colleagues who made the program feasible through their contribution to the research agenda, reviewing proposals, discussing interim and final results, and interacting with CIOs and sponsors of the program. I would like to express my deep gratitude to colleagues who participated in the different stages of the program: Jean-Eric Aubert, international consultant and former lead specialist at the World Bank, Surinder Batra, Institute of Management Technology, Ghaziabad, Michel Beadouin-Lafon, University Paris-Sud, Pierre-Jean Benghozi, Ecole Polytechnique, Marcos Cavalcanti, Federal University of Rio Janeiro, Leif Edvinsson, University of Lund, Patrick Fridenson, Ecole des Hautes Etudes en Sciences Sociales, Dominique Guellec, OECD, Thomas Housel, Naval Post-Graduate School, Junichi IIjima, Tokyo Institute of Technology, Moez Limayem, University of Florida, Rik Maes, University of Amsterdam, M. Lynne Markus, Bentley University, Peter Meusburger, Heidelberg University, Ian Miles, University of Manchester, Yves Pigneur, University of Lausanne, Frantz Rowe, University of Nantes, Gérald Santucci, European Commission, DG Connect, Pirjo Stahle, Aalto University, and Eric Tsui, Hong Kong Polytechnic University.

Thank you also to the 50 teams around the world who contributed to the program's workpackages and other activities.

I wish especially to thank Anne-Sophie Boisard, Mission Director, CIGREF, the linchpin of the whole program, who patiently managed its tasks and deliverables on a daily basis.

Finally, warm thanks to my editor, Christian Rauscher, who kindly and constantly supports ISD progam publications, especially via SpringerBriefs in Digital Spaces, as well as to Barbara Bethke who, as usual, took care professionally of the organizational tasks related to our collective editorial effort.

Contents

Chapter 1
Introduction

This chapter presents the overall rationale for the book and its positioning. It defines the framework for the Information Systems Dynamics (ISD) research program that was sponsored by CIGREF and other large international companies. It establishes the context for digital transformation and organizational design, explains why the program is focused on organizational design, and how it relates to the use of IT in companies and societies.

This book is about the current and future digital transformation of firms and organizations. This is an issue that affects managers around the world, evidenced by the growing number of workshops and seminars that address the inherent risks and potential. While there are clearly risks, expectations are high, particularly with respect to the value to be extracted from one of the major assets of digital transformation: data.

Digital transformation is also an issue for society, as digital ubiquity affects everyone in their daily lives. Not only business models, but also society and families are deeply and constantly challenged by their everyday interactions with technology.

There is a need to go beyond a general discussion and understand the nature of digital transformation, in order to design plausible futures for digitality in firms and organizations. That is the main aim of this book, which takes a broad interdisciplinary approach to a complex phenomenon that is still emerging. In addition to understanding the phenomenon itself, we also need to identify its underlying production mode and governance structure.

This book is based on the results of the international ISD research program led by the CIGREF Foundation. The program ran from 2008 to 2014, and supported around 30 projects and 50 teams worldwide (the United States, Europe, and Asia). It provides an integrated view of key trends in digital transformation from five interrelated perspectives: strategic (business models), societal, organizational, technological and regulatory.

© Springer International Publishing Switzerland 2016
A. Bounfour, *Digital Futures, Digital Transformation,*
Progress in IS, DOI 10.1007/978-3-319-23279-9_1

This book complements other ISD deliverables:

- The research agenda[1]
- An overview of Wave A projects[2]
- Acceluction II: An overview of Wave B projects[3]
- Two specialized collections published by Springer: Espaces Numériques[4] and SpringerBriefs in Digital Spaces[5]
- Work summarized in the ESSENTIALS collection
- The Wave C report

The book provides a framework for the analysis of digital business transformation in terms of emerging factors. It examines 25 key future trends. Based on these trends, the book delineates a new mode of production, centered on accelerating the links between multiple value creation spaces (customers, suppliers, data, complementors, social networks, etc.). This proposed mode of production—*acceluction*—is analyzed in terms of its rationale, content and implications for the management of firms, especially executives who are responsible for the digital transformation agenda.

Acceluction is analyzed through a comparison with the current, dominant production mode, Lean production, developed by the International Motor Vehicle Program (IMVP) in the early 1990s. The new production mode is also described in terms of one of its intrinsic aspects: the plurality of tensions. Tomorrow's management, supported by digital transformation, reflects many different tensions; notably between internal and external resources, horizontality and verticality in organizations, and short timeframes for decision making, all driven by the acceleration and long-term perspectives needed to build resilient enterprises.

The implications of this analysis are explored in terms of governance, taking into account six proposed scenarios. Finally, the potential post-2020 situation is examined in terms of the expected impact of an abundance of networks and data, which may become critical resources for the enterprise.

1.1 ISD as an International Research Program

ISD is a public interest program that aims to evaluate the societal and managerial challenges in the long-term usage of IS (1970–2020) by mobilising the best expertise available at the international level. Its purpose is to analyse the momentum of usage by studying its different dimensions: strategic, social, ethical,

---

[1]Bounfour (2010a).

[2]Bounfour (2011).

[3]Bounfour (2013).

[4]http://www.springer.com/series/11459.

[5]http://www.springer.com/series/10461.

regulatory, technological and organisational—both from a historical point of view and from a prospective one.

ISD considers that the issue of use of information systems and digital transformation now goes beyond the framework of managerial action: it embraces the whole domain of society.

The ISD international research program was initiated in 2008, with four main objectives:

- To link the past with the present—the "long view" of history, in which we can see general trends and regularities—with the "short view" of the life of the company. Even if the historical dimension is not central to the ISD initiative, the program has backed several studies examining the history of information system uses, some of which were published in a special issue of the review "Entreprises et Histoire" in December 2010, to mark the fortieth anniversary of CIGREF. Other programs were also touched upon that addressed the question of the history of usage and its impact on business design and performance, notably the MIS project at Harvard and the Japanese program sponsored by METI, which we discuss later.
- *To alert the leaders of large companies to the strategic implications of this sea change: the transition from an industrial economy to a networked economy based on knowledge and immateriality.* This raises two questions: (1) How—by what means—will digital bring about a paradigm shift in businesses' "modes of production" of value? and (2) How does this paradigm shift intermesh with the development of immateriality and its component parts (contracting modalities, reputation effects, changes in intellectual property rights, the time dimension of performance, etc.)? Is digital an independent driver of business transformation or is it an accelerating factor? These are basic analytical questions: questions that the research must address, looking beyond the handful of "case studies" repeatedly alluded to—often without any supporting arguments—in the general discourse on digital transformation.
- *To propose basic building blocks for an analytical grid, justified by research findings from the program and other available elements.* By "analytical grid" we mean the set of basic conceptual elements that structure the thinking and action of enterprises, and more widely of all public and private policy-makers concerned by the impact of digital transformation on business. The questions here are simple. They concern the way in which economic activities are now organized in the program's primary economic field: the enterprise. The accent is therefore on organizing, i.e. on the way the activities of enterprises are articulated in time and space, and consequently on their control and governance.
- *To shed light on the future of the "Information Systems" function, by drawing on studies of emergent factors and modeling.* Information systems represent a major share of companies' investments (25–30 % of total investment in some cases). They are also an essential value creation lever for customers and for the entire ecosystem of enterprises and public and private organizations. But when it comes to information systems (IS)—or indeed any investment, whether tangible

or intangible—analytical reasoning has been largely dominated by unitary reasoning, confined to the enterprise and its more or less well-defined boundaries (the persistence of thinking in terms of value chains is a case in point). How does the ongoing digital transformation change the rules of the game? Will it lead to a differentiated positioning for the IS function? How will digital governance be exercised in the enterprise of the year 2020? What will the associated functional profiles be? These are just some of the questions on the agenda of companies' leaders and their internal and external stakeholders.

In light of these objectives, the program research agenda proposed five analytical perspectives as the core building blocks of the programme. The future of enterprises—and the design of their future IS—will be determined by the interaction between developments in socio-ethical, strategic, technological, regulatory and organisational trends.

It is by considering these five perspectives, interactively and systemically, that we can grasp the reality of the driving forces affecting future companies and their information systems.

- **Social and ethical trends**, because it is by analysing these trends within societies and their evolution—including in terms of innovation and IT—that future norms of conduct will be brought to the fore;
- **Strategic trends**, because today's business models inevitably face major disruptive change, in which networks, communities and spot markets will play a decisive role;
- **Technological trends**, because the ubiquity of informational artefacts will become commonplace;
- **Regulatory trends,** because new standards—including regulation—will emerge to regulate the transitions that are taking place (we must therefore be prepared for and, of course, influence such tendencies);
- **Organisational trends**, because companies and societies will undergo massive transformations in their structures, processes and standards.

The ISD research program is therefore focused on obtaining a better understanding of the relationship between, on the one hand, progress in the development and implementation of information technology, and information systems and on the other, their numerous impacts on organisations, industries, and society in general. This relentless progress is building a future where digitality will play an increasingly important role in:

- Business models;
- Organisational design;
- Support for innovation;
- Human resource strategy;
- Governmental regulation;
- Social and economic development.

A critical enabler of these profound changes will be the existence of a ubiquitous broadband network infrastructure (wireless and wired), not just in mature economies but also in developing nations. The impact of the evolving IT superstructure on societies, and the ethical practices they embrace, will also have to be considered as a factor in IT/IS leadership.

The very concepts of time, space, and information complexity will have to be revised to help explain and predict the trajectories of these changes. This will ultimately affect the management of IT capabilities by identifying emerging practices embodied in performance rules and developing standards.

1.2 Business Models and Digitality

Business models are closely related to value creation and its modalities in the knowledge economy. Value creation has been widely debated in the economics and business literature over the past ten years. Since the advent of the Internet, several—sometimes naïve—views have been put forward, notably those that emphasize the role of transparency and resource mobility. In many respects, the new economy is considered to be endowed with characteristics that are similar to the market structures proposed by classical economists, notably related to atomicity, and the free entry and exit of firms in the market. The main issue is the extent to which the new economy fosters specific characteristics, which require (or include) a profound change in business modelling from the industrial point of view (the value-added chain). Are there alternative or complementary ways to design business models in the new economy? Has the co-production of value become an alternative to linear production? This is a distinction that was introduced by Ramirez (1999), as an extension to his value constellation model. Ramirez differentiates between production value in value-added chains (basically an industrial perspective), and value co-production (a model that is more adapted to the intangible economy).

From the industrial point of view, value creation is sequential, unidirectional and transitive. It is realised through transactions and measured in monetary terms. Economic actors have a dominant role in a linear sequence of value-added chains. From a co-production perspective, "value creation is synchronic, interactive, best described in 'value constellations'" (Ramirez 1999: 61). Consumers are considered as a factor of production and contribute to value creation. Economic actors hold different roles simultaneously, and the basic unit of analysis is the interaction.

Underlying Ramirez's analysis is the assumption that we are facing a deep shift in business paradigms, where value is created by co-production rather than sequentially in linear terms. In this paradigm, interstices have more relevance than sequences. From the business perspective, value creation is linked to the intangible, relationship dimension of activities (Baxter and Matear 2004). This raises the question: is Porter's value-added chain (Porter 1985) still relevant?

The foregoing underlines that in value-added chains, value is created through sequential processes that transform inputs into products and rely on two types of activity: primary and support. In the 'Value Shop' model the basic logic of value creation lies in mobilizing resources towards problem-solving. Value is created for customers by solving their problems. In the 'Value Network' model, value is created by linking independent customers, directly or indirectly (Stabell and Fjeldstad, 1998). This distinction highlights the importance of taking a broader view of the value-added chain, in particular through the consideration of business combinations and hybrid market structures that might be created by the new economy, notably from a B2B perspective.

A detailed assessment of hybrid forms of organizing and the societal dimension of digital use (including time-space), makes it clear that the research agenda for business modelling needs a profound re-evaluation. All the more so given the development of societal and platformic innovations around what is now called the "positive" or "shared" (access) economy. These developments, which have been facilitated and even shaped by digital resources, require a fundamental re-examination of thinking on business modelling that goes beyond the firm's traditional boundaries and value chain.

1.3 ISD and Organisational Design

The aim of Information Systems Dynamics (ISD) is to identify critical signals related to emerging digital uses and then, based on these signals, to design the digital uses, value creation spaces, and mode of governance of the 2020 enterprise.

The ISD research agenda was focused on the theme of organizing; in other words, how businesses—large enterprises in particular—should articulate their material and digital resources as we approach the year 2020. Here we focus on strategic, organizational, and social dimensions, rather than technological or regulatory.

A central issue for the conceptual framework set out in the research agenda was the organizational design of the enterprise. By organizational design, we mean the *identification and interplay of the conceptual building blocks* that help in understanding the enterprise, its digital uses and associated governance. Naturally, as the program is the initiative of large companies, 'enterprise' should be understood as referring to an organization with resources that exceed a certain significance threshold. Specifically, we focus on organizations that traditionally constitute a 'center' for economic production at the national and international level, but that are also a center of influence in terms of standards, practices and managerial discourse. This factor is unrelated to the program's sponsors; rather it is a fundamental question that goes beyond one specific criterion. No-one would deny that over the past sixty years the trajectory and, to some extent, the transformation of socio-economic systems have been significantly influenced by large companies. The current question managers of (large) companies and their stakeholders (public policy-makers, analysts, investors) face is how to respond to the ***paradigm shift in modes of production*** brought about by digitality. One of the consequences, and fundamental challenges, of this shift is a rearrangement of value creation spaces, in which certain essential components may move beyond the traditional frontiers of the large company, and thus beyond its control.

The core objective of the ISD research program is to understand emerging (and ongoing) organizational and digital transformations, with the aim of identifying their basic conceptual building blocks.

These two dimensions provide the focus for the analytical work carried out by research teams supported by the program, and the discussions of the program's governance bodies: the Steering Committee, the Scientific Committee, the Strategic Orientation Committee, and the Organization Committee.

The result is a unique procedure for the production of collective knowledge about one of the great transformations at work in the 21st century: the digital transformation.

1.3.1 Organizational Design: An Issue for Renewal

Digitality and the various layers of transformation highlight the issue of organizational design as a topic for researchers and practitioners, particularly in the domains of organizational science and economics. In this context, organizational design refers to how firms conceptualize and effectively articulate tangible, financial and intangible resources. In the domain of organizational economics, the issue has been considered in terms of the firm's boundaries: transactional cost economics has developed an elegant framework for understanding the trade-off between markets and hierarchies (Coase 1937; Williamson 1975, 1981). On the other hand, organizational science explains the existence of organizations in terms of their fundaments, notably the importance of aggregating and developing skills in a very specific organizational setting (Chandler 1992), the reduction of uncertainty stemming from the environment and technology, and more generally, information complexity (March and Simon 1958; Thompson 1967).

This shows the two approaches to organizational design: on the hand, organizational boundaries depend on optimizing transactions, and on the other hand, organizations exist because of their intrinsic characteristics—idiosyncratic skills and administrative processes put in place to deal with uncertainties and information complexity.

When examining information technology in particular, approaches that are framed in terms of an organizing vision (Swanson and Ramiller 1997) provide a framework for understanding the deployment of digital artifacts within and around organizations. There are three dimensions that should be considered in order to understand such a deployment: interpretation, legitimation and mobilization.

1.3.2 The Future of Organizing—Beyond Web 2.0 Organizations

McAfee (2009) defined and popularized the concept of the web 2.0 enterprise. It basically refers to how enterprises leverage social tools to achieve their objectives, "Enterprise 2.0 is the use of emergent social software platforms by organizations in pursuit of their goals" (McAfee 2009: 73). From this perspective, Enterprise 2.0 is not a technological phenomenon but is instead related to norms, policies and guidelines, illustrated by the example of Wikipedia.

However, there are serious arguments that go beyond the traditional two approaches to organizational design that have been developed by economists, which call for a fresh view that is based on emerging practices. From a more holistic perspective, three issues illustrate the need to renew the organizational design agenda: organizational architecture, open innovation, and digital space.

1.3.3 Organizational Architecture

The issue of organizational architecture is closely related to the digital transformation and its impact on digital strategies. A special issue of the *Strategic Management Journal* (June 2012, Vol. 33, No. 6) was dedicated to strategy and the design of organizational architecture, with a focus of topics such as networks, organizational structures, attention and adaptation. In this issue, Gulati, Puranam and Tushman (2012) emphasized the importance of the design of the meta-organization as a perspective for future research.

Organizational architecture (and its building blocks) is a current issue due to the extreme plasticity of organizations both in terms of space (where the question of the organizational boundary is one aspect) and time (the acceleration of everything). Therefore, organizations need to develop innovative capabilities for the ongoing design of their resource architecture, concomitant with the operational capacity to rapidly take advantage of such capabilities. In other words, organizations need a new type of organizational capital that specifically addresses the articulation of these two capabilities.

1.3.4 Open Innovation

Open innovation has become an important topic for managers (Chesbrough 2006; Chesbrough et al. 2006) in both theory and practice. In terms of organizational design, this means that the focus for innovation is moving from the organization to the different forms of external socio-economising spaces (spot markets, more-or-less organized communities). It is clear that this change impacts the organizational agenda for firms, and how they allocate resources. The digital component of organizational design is very important. Open innovation has been facilitated by reduced communication costs that have led to the development of market-driven sources of innovation (Tushman et al. 2012). As a result of open innovation, we posit that firms and organizations are becoming less and less centric, and therefore increasingly boundaryless.

1.3.5   Digital Space and Data

The ubiquity of digitality and the resulting massive explosion of data are challenging the way firms and organizations can benefit. The issue of 'big data' is important because of its quantity, lack of structure, and real-time nature. Organizations that want to take advantage of big data need to implement embedded processes, which creates a power shift and therefore resistance to change. Galbraith (2014: 3) quotes the case of Procter & Gamble, which established an analytical group in 1992; this eased the adoption of big data analytics and led to the development of new skills. Implementing the same procedures in other contexts can lead to problems as it can create a shift in power between specific functions or activities, although this may be exactly what is needed to embed data analytics into organizational processes. Organizational culture and processes are therefore important elements to take into account when considering firms' readiness. One solution may lie in the transformation of organizations, from command-and-control hierarchies towards self-organizing structures that empower staff at all levels (Graupner and Maedche 2014). Another option might consist in considering data as a transversal issue that is seen as a core asset for the whole organization.

Table 1.1 Dimensions of organizational design and digitality

Dimensions of organizational design	Questions for managers	Topics for reflection
Digital ubiquity	• How to develop a transversal view of digitality that goes beyond functional silos? • How to take advantage of ubiquity in order to develop innovation?	Innovation strategies in the context of digital ubiquity
Open organizations	• How to articulate "internal" and "external" resources? • What specific processes and incentives need to be deployed?	Designing incentive systems in the context of open organizations
Data as assets	• According to what criteria can data be considered as an asset? • How can data be valued as a digital asset?	Developing approaches to valuing digital assets

1.4   Organizational Design: Questions and Dimensions

The issues discussed above raise specific questions for managers, and suggest
topics for further reflection (Table 1.1). Digital ubiquity, open organizations, and
data are three critical dimensions that need to be given careful consideration by
managers. Based on the design of organizations, they call for a renewed research
and action agenda, which goes beyond the traditional functional silo approach.

Chapter 2
From IT to Digital Transformation: A Long Term Perspective

This chapter takes a long term perspective for the use of IT artifacts in companies and organizations. It analyses the semantical shift in the use of digital artifacts, from IT to IS to digitality. It also reviews how IT use has been integrated in the organizational design and the specific role of "organizational fit". The chapter also reviews three major international programs which developed a historical perspective for the use of IT: the US program (Harvard), the Japanese program (supported by METI) and the French program, as supported by CIGREF. Finally, the chapter justifies how the concept of use, control and innovation are now shifting from companies towards the whole society, and the specific role of social media as new spaces for innovation and appropriation of IT artifacts.

2.1 Historical Perspective[1]

The historical perspective is useful here; as it puts the questions facing business leaders—and the ways information systems are used—into context.

It is instructive to look back at some of the studies and programs that have tried to outline a history of corporate computerization and, in the process, to reveal how information systems and technologies have been used strategically and operationally in given contexts. Such works are still fairly scarce. We will review some of them briefly here, first and foremost the Harvard program.

2.1.1 The Harvard MIS History Project

One of the first historical research papers on the management of information systems was published by Copeland and McKenney in 1988.[2] This paper, which

---

[1] In this section I pick up on some of the ideas developed in the introduction to the special issue of *Entreprises et Histoires*, no. 60, "De l'informatisation aux systèmes d'information dans les grandes entreprises", 2010.
[2] Copeland and McKenney (1988).

© Springer International Publishing Switzerland 2016
A. Bounfour, *Digital Futures, Digital Transformation*,
Progress in IS, DOI 10.1007/978-3-319-23279-9_2

studied air transport by means of historical analysis, revealed the importance of sector factors (deregulation), but also of factors specific to airlines in gaining competitive advantage via their reservations systems: as well as economies of scale, (cumulative) technological experience played a crucial role in establishing automated booking systems. In 1976, the authors tell us, the major US carriers (American, United, and to a lesser degree TWA) had stable distribution systems, and innovations could only be introduced incrementally. But these factors were insufficient; the managerial vision of the groups' leaders played a determining role. Several factors emerge as critical to the successful deployment of a reservation system by airlines: economies of scale, and what the authors call "intelligent persistence" (p. 368), in other words a practical vision—particularly by business leaders—of how to use technology.

This article opened the way for developing the conceptualization of the historical method in information systems research, around the Harvard project. The project was launched in 1988, to work on analyzing the impact of IT investments on business, industry and society. Coordinated by Harvard Business School professors James L. McKenney and John G. McLean, it involved a project team of nearly a dozen researchers and experts. The research focused mainly on analyzing the impact of information technology investments on four large companies, on which detailed case studies were conducted, including Bank of America and American Airlines. The research concluded that there was a need to develop a historical method to gauge the impacts of information technologies on enterprises. This was the subject of three papers published in *MIS Quarterly* in 1997 (no. 3) by the main project contributors.

The first, written by Mason et al. (1997b) raises the issue of the place and importance of historical research as a component of the information systems discipline and its general epistemological framework. It sees historical research as a factor in the legitimation of information systems research. After underlining the important difference between the history of information systems management and the history of information technologies, the authors formulate a general conceptual structure for what they call "IT-based business histories". Based on the four case studies analyzed—following the familiar Harvard method—the authors seek to understand how certain organizations manage to take advantage of information technologies to establish a dominant competitive position. Their answer is a modeling of companies' information technology uses by means of a "cascade" model, in which the ultimate phase is the imposition by the firm of a "dominant design" of information system uses. A dominant design is defined here as a radical design of a product or service liable to bring about a major change in the competitive landscape of an industry (the DC3 and the Ford T are cited as examples). The dominant design is the culmination of a cascading process that requires three roles to be performed: the roles of *leader*, *maestro* and *supertech*. The leader is an actor within the company, who deploys leadership resources to leverage the potential of technology; the example used is that of S. Clarke Beise, appointed President of Bank of America in 1954. The maestro is an IT manager with a thorough grasp of the company's business as well as technology (still a hot topic in

IS departments). Finally, supertechs are members of the project team itself, operating with a degree of harmony. The authors propose a five-phase cascading model, centered on these three roles. Starting out from a crisis situation, the company's actors look for a technical solution (Phase 1), then initiate the solution (Phase 2), adjust the organizational structure (Phase 3), and train assets who solve the problem, with the aim of gaining a competitive advantage (Phase 4), in order to assert a dominant design (Phase 5).

This conceptual model is used in the second article by McKenney et al. (1997) where it is applied to the case of Bank of America, demonstrating how the bank used information technologies as a crucial lever for attaining sector leadership in the USA—and the world in general—in the 1950s. The bank fundamentally changed the nature of the banking industry by deploying the ERMA and IBM 702 systems, which enabled it to develop a dominant IT design based notably on an articulation of three roles essential to the emergence of such a design: Clark Beise (leader), Al Zipf (maestro) and a group of supertechs. The cascading model proved its effectiveness here: starting from a crisis situation (IBM's inability to deliver a fully operational system for its 360/65), it became possible to produce a dominant design, enabling the bank to assert its leadership for a decade and a half.

The third article by the same three authors[3] proposes a seven-steps methodological framework for the conduct of historical research on information systems management.

2.1.2 The Work of Chandler and Cortada

In parallel to the Harvard managerial project, other American studies focused on the question of the history of information system and technology uses and their impact on business and industrial performance, most notably the work codirected by Alfred Chandler and James Cortada, followed by the three volumes published by James Cortada, dealing with the transformation of the American economy, and of American industries and administrations, by information technologies.

The collection directed by Chandler and Cortada (2000) offers a long-term perspective on the transformation of the American economy by information—and the establishment of informational infrastructure and interactions in the USA— since the 1700s. The work of James Cortada, who was also one of the contributors, gave rise to three important volumes, in which the impact of the "digital hand" (by analogy with Adam Smith's invisible hand) on industry and on the administration is analyzed in detail.

The first volume[4] looks at the role of information in the transformation of manufacturing, transportation and retail distribution, from the 1950s onward. Uses

---

[3]Mason et al. (1997a).
[4]Cortada (2004).

are analyzed sector by sector, but also transversally, highlighting the importance of imitation in digital uses (as with the PC, of which the spread was induced by its adoption by IBM in 1981). It also underlines the facility with which information system applications were adopted by most industries, and how these radically changed the way industries operated, by transferring whole swathes of manual tasks to computers. This rapid spread was facilitated by early adopters: large companies to begin with, followed by smaller firms, and by initiatives taken within particular industries. Standardization, and the availability of commercial software, also played a determining role.

What managerial lessons can be drawn from this analysis? There was a clear interest in IT in the 1950s, but managers were very attentive to the economic aspects of the related investment, in a context characterized by senior management's poor understanding of the technology's potential until the 1970s. A fundamental change was brought about in the mid-1980s with the democratization of the PC. Suddenly, information systems were a strategic question.

The same structure is used in Volume 2,[5] which centers on the transformation of financial services, telecommunications, leisure and the media. The analysis of these branches throws up a number of specific problems, in particular those relating to intellectual property rights, as well as specificities in the structuring of information artifacts (such as databases in banking).

The third and final volume[6] deals with the public sector (defense, education, fiscal services, etc.). It particularly insists on the role of the defense sector (through government contracts) in creating a competitive IT industry.

In her book *structuring the Information Age*,[7] JoAnne Yates examines the adoption and use of information technologies by American insurance firms during the 20th century. She reveals the influence of uses on the evolution of information technologies, and vice versa, notably demonstrating how the use of tabulators in insurance influenced the adoption and use of the first computers, and how these computers had a similar relationship with the technology and the firms that produced them.

This American perspective underscores the importance of information system uses in the transformation of organizations and economies. In the works of the Harvard project and in those of Chandler and Cortada, the central question is that of usage, and how it transforms organizational and individual practices. At the level of the enterprise, the question concerns the definition of a "dominant design" that will produce a rent for the enterprise, and thus a lasting competitive advantage. At the macroeconomic level, the central questions are job creation, economic modernity and the technological leadership of a country or collectivity (in this case the USA since the 1950s). At the social level, the question can be framed in terms of the transformation of interpersonal relations by information artifacts and information

---

[5]Cortada (2006).

[6]Cortada (2008).

[7]Yates (2005).

systems. The links to the Giddensian perspective and the theory of structuration are easy to see, as is the influence of the boundary-object reasoning advocated by Star and Griesemer.

2.1.3   The Japanese Initiatives

A number of Japanese studies on the history of digital uses were carried out in the 1980s, albeit with limited results. These often little-known works are described in Kiyoshi Murata's paper (2010). One of the first pieces of research on this topic was published by the Information Processing Society of Japan's Special Committee for the History of Computing in 1985 and 1998. Several projects focused on the contribution of information systems to organizational change, business performance and distribution of tasks between actors. But the work of particular interest to us here is the joint project launched in April 2007 by the Japan Society for Management Information (JASMIN), the Ministry of Economy, Trade and Industry (METI), the Japan Information Processing Development Corporation (JIPDEC), the Japan Users Association of Information Systems (JUAS) and Nikkei Business Publications. The project's goal, which involved some ten researchers over 2 years, was to gather in-depth knowledge about the conditions of use of information systems, both at the level of the overall economy and by sector (manufacturing, retail and financial services).

Unlike the Harvard project, with its strong focus on business and on competitive positioning, the Japanese program has a more systemic orientation, due to the plurality of actors; actors with an interest in how companies—and large corporations in particular—are transformed by information technologies. Through a collective research effort, these actors are able to gain insights into their respective involvement in deploying information system and technology uses in their organizations, and the factors behind their deployment, for example in terms of the balance of power between the different stakeholders, and their modalities of cooperation.

2.1.4   Research in France and the ISD Research Program

Historical research in the field of managerial information system uses is a recent phenomenon. It can, for the most part, be tied in with work by management scientists and historians. But overall, it is still an emerging area of research.

CIGREF initiated one of the first studies on the changing position of the information systems function in large companies.[8] This research, by Ravidat,

---

[8]Ravidat et al. (2005), Ravidat and Akoka (2006).

Schmitt and Akoka, presented a longitudinal analysis of the evolution of the position of the IS function from 1992 to 2004, collating 12 years of publications, including 146 reports. The resulting analytical model brings five components into play:

- The 5 main actors of the information systems function (the business units, the users, the suppliers, the integrators, and branch IT);
- The technical and economic environment;
- The points of balance or compromise between the 5 actors;
- The core business of the IS function;
- The main distinguishing features identified.

This longitudinal study of the key features and interactions of these five elements made it possible to characterize the evolution of the IS function's position within organizations. A notable observation was that the explosion in the supply of technological products was accompanied by an increase in the responsibility of the IS department and its consequent reorientation towards less technical issues, as from 2000. More generally, the study identified key periods that characterize the position of the IS function in the enterprise, and a characteristic line of development of its "core business" (Table 2.1).

The historical research in France also looked at the agenda of information systems researchers over more than 25 years. Like their Anglo-Saxon colleagues before them,[9] Desq et al. (2002, 2007) analyzed the topics studied by information systems researchers, bringing out the specificity of Francophone research, notably in terms of methodology and the central themes developed. Similarly, the work of Nathalie Greenan and her colleagues[10] on the evolving relationship between ICT investments and new organizational practices in French companies, which draws on the COI survey, has an obvious historical dimension, revealing as it does a strong positive correlation between ICT uses and the quantity of new organizational practices during the 1990s, with moments of acceleration in company reorganizations running parallel to growth in their ICT investments.

2.1.5 The ISD Program

The ISD program is built around a central concept: organizational design, bearing in mind that deconstruction is all around us, in business, society and international power struggles. This concept of design is not new. Looking no further than the sphere of information systems, one of the recurrent questions in the literature is that of the development of an *organizing vision* of innovation.[11] But it has been put

---

[9]See notably Culnan and Swanson (1986).
[10]Greenan and Mairesse (2006), Greenan et al. (2010).
[11]Swanson and Ramiller (1997), Ramiller and Swanson (2003).

Table 2.1 Evolution of the IS function in large French companies (1992–2004)

	Core competence	Positioning
1992–1994	Coherence and control	Technical positioning
1995–1996	Architecture, service continuity	Technical and services positioning
1997–1999	Security, cost, service	Loss of technical control, positioning as internal software house
2000–2002	Knowledge management, governance	Positioning towards the strategy
2002—2004	Urbanization, certification, business intelligence, legal	Search for new legitimacy

Source Ravidat, Schmitt, Akoka (2005: 7), translated by the author

back center stage by the fundamental changes under way in technology and in the structure of businesses and business activities, and the emergence of new organizational models in which immateriality plays a key role.

In the case of ISD, organizational design is a conceptual tool that should enable us to determine the core characteristics of *organizing 2020*, starting from the five dimensions developed earlier, and paying close attention to emerging factors (along the lines of the MIT program on the future of the automotive industry in the 1980s). The program posited that growth in the use of artifacts and information spaces (via digital), and their increasing ubiquity, associated with a massive transformation of modes of production and social interaction, including in the workplace, and the arrival of emerging actors, will generate new organizational configurations in the enterprise—configurations that need to be characterized. If digitality is a structural given of modes of production and interaction, it is also clear that the future modalities of interaction will in part be determined by managerial practices and modes of contractualization at the societal level. Will the liquid society defined by Bauman (2000) find its counterpart in the "liquid enterprise"? As we will see, liquidity is one of the key dimensions in the design of the 2020 enterprise.

The special issue of *Entreprises et Histoire*, which marks CIGREF's 40th anniversary, considers the question of the IS-driven transformation of businesses from a historical perspective. Some of the research published here refers to work conducted and defended during the prototyping phase of the ISD program, or aligned with the program. In keeping with the general stance of *Entreprises et Histoire*, this issue presents research conducted by management experts and business historians in France and in two countries where thinking on the information-system-driven transformation of businesses, from a historical perspective, has been particularly developed namely the USA and Japan. It also reports on an interesting and stimulating debate between CIOs, CEOs and researchers.

The paper by M. Lynne Markus (2010) that opens the issue raises the question of how the use of information technologies articulates with the design of organizations. It reviews the long term evolutions of US large corporation since the M form deeply analysed by the work by Chandler. The paper analyzed the success of organizational design during the last period: from the IBM (mainframe) era to Microsoft (client-server), to probably the now google era. The paper insists on the co-evolutionary character of IT technology and organization design. It also underlines the need to develop a specific research agenda on the topic of organizational design.

I have already stressed the importance of James Cortada's contributions to the analysis of the impact of information technologies on businesses and industries in the USA. His article discusses the importance of industries in analyzing the impact of information technologies, but also for the professional positioning of information system managers. According to him, industries constitute an important scope for analyzing the transformation of firms by IT (Cortada, 2010).

Kiyoshi Murata's paper (2010) reports on the lessons learned from the analysis of the development of information systems in Japanese companies. This research points to a disruption in the triangular balance between large companies, large administrations (METI) and large service providers to the benefit of the latter, due to the spread of outsourcing practices. These practices have weakened companies' capacity to absorb and to transform, because they have lost the necessary skills. A managerial lesson that European businesses would do well to reflect upon.

Alexandre Giandou's text (2010) has an institutional scope. It describes the conditions for the creation of CIGREF in 1970, the end purposes defined by the founders, and the institutional credibility that CIGREF has since acquired in the field of information systems.

Rodhain et al. (2010), meanwhile, have compiled a history of information systems research based on the analysis of 1945 articles published between 1977 and 2008. Their analysis breaks the development of management information system research down into key periods, notably with regard to the dominant research topics and methodologies. Management information system (MIS) research is still a young discipline, but it has already demonstrated a capacity for theorizing and for developing a specific agenda, in phase with the concerns of businesses and society.

Griset's article (2010) analyzes the first period in the adoption of information technologies by enterprises in France. In it, we discover that the introduction of the computer raised more problems than it solved, and called for a fundamental rethink of organizational and informational mechanisms and processes. The logic of flows came into its own in the early 1970s. The concepts of "network" and of "real time" are major vectors of computerization.

In a complementary paper, Alain Beltran[12] studies the arrival of IT and the organization of French companies from the end of the 1960s to the beginning of the 1980s. His research sheds light on the organizational dimension of business computerization, the structuring of tasks, the articulation between centralization and decentralization, and the democratization of IT knowledge beyond the circle of specialist technicians.

---

[12]Beltran (2010).

François Hochereau[13] offers an in-depth analysis of the transition to computer-ization of a large telecommunications company, by reconstituting "successive organizing visions of a strategic business process". The different stages in com-puterization are highlighted, notably with a customer orientation observable in other enterprises. The article confirms the validity of thinking in terms of an "organizing vision" of technological innovation driven by the information systems, but it also describes its limitations, in particular its inability to integrate spiral developments, which reflect contradictions that can be resolved through recourse to other technical innovations.

The article by Pierre-Eric Mounier-Kuhn[14] describes the role of user clubs in the evolution of products and practices, around particular vendors and more generally in the dissemination of knowledge between users, and in defending IT investments to stakeholders (investors, government agencies).

In this special issue, the discussion between CIOs and researchers is interesting (p. 70–84). It provides an overall vision of a history of information systems and technology uses in business, and highlights a number of discussion points, partic-ularly as relates to the role of government in real-life transformation processes.

2.2 The Long-Term Perspective

To understand the role of the digital transformation, it is important to consider it from a long-term perspective. As Malecki and Moriset (2008) underline, the digital revolution can be considered from the perspective of the Kondratiev cycle (Malecki and Moriset 2008: 26). This approach has been analyzed by several scholars (Atkinson 2004; Freeman and Louça 2001; Louça 2003). For example, Freeman and Louça (2001) analyzed successive industrial revolutions, notably: (1) the British industrial revolution (the age of cotton, iron and water power); (2) the second Kondratiev wave (the age of railways, steam power, and mechanization); (3) the third Kondratiev wave (the age of steel, heavy engineering, and electrifi-cation); (4) The Fourth Kondratiev wave (the great depression and the age of oil, automobiles, motorization, and mass production); and (5) what authors have named a new techno-economic paradigm, the age of information and communication technologies. This paradigm is characterized by the widespread use of computers and telecommunications, and is accompanied by an organizational change, namely the "network firm" (Freeman and Louça 2001: 324–327). It represents an institu-tional change, new modes of regulation and a culture of virtuality.

The current information revolution is based on digital, entrepreneurial and knowledge-based discovery, together with a change in managerial practices and organizational design (Malecki and Moriset 2008). These three ingredients are the

---

[13]Hochereau (2010).

[14]Mounier-Kuhn (2010).

key to understanding the ongoing digital transformation and raise the question: How do managerial practices and entrepreneurial discovery interact with digital artifacts to build a new production system?

There is a common thread that emerges from this brief overview: the articulation between information systems and organizational design. Digitality, as a system, changes the game. It amplifies the impacts that researchers have identified, due to its ubiquitous and transformational nature that goes beyond the traditional boundaries of the enterprise. Therefore, one of the expected outputs of the ISD research program is to answer the question: How does digitality articulate with the design of enterprises, and how does it affect their transformation?

2.3 Digital Transformation

Digital transformation is a new development in the use of digital artifacts, systems and symbols within and around organizations. Although the term does not have a clear definition, it encompasses several dimensions. Recent McKinsey reports have examined the issue of digital transformation via the impact of the Internet. Several reports have been issued that target both businesses and policymakers.

Bughin and Manyika (2012) provide an analysis of the impact of the Internet on economic growth at a national scale, given the respective positioning and potential of countries. They found that the Internet contributed 3.4 % to GDP in the 13 countries they looked at. The United States leads the supply ecosystem. In Europe, the United Kingdom and Sweden are game-changers, while France and Germany are very influential in terms of usage. The position of India and China is becoming stronger, while two of the Brics (Brazil and Russia), together with Italy are still in the early stages. More recently, Woetzel et al. (2014) report on the importance of Internet use by Chinese citizens. Some key figures include: 632 million Chinese Internet users; 700 million active users of smart devices, and US$ 300 billion earned in e-tailing sales in 2013. Internet use in China tends to be consumer-oriented and the market share of the Internet-oriented economy is 4.4 % GDP, which is higher than in the United States or Germany. In France, the 2014 McKinsey report (McKinsey and Company 2014) underlines the disequilibrium between widespread consumer user and business use. The latter appears to be lagging behind due to a lack of skills, financial constraints and commitment from senior management. This suggests that the digital transformation is a process that involves the diffusion of the Internet in both demand and supply sectors. It therefore differentiates performance between countries in terms of the level and extent of the use of Internet applications. This use reflects the level of maturity of countries and disequilibrium between demand and supply.

In the same vein, but from a microeconomic perspective, Cap Gemini and MIT Sloan Management (2011) examined the issue of digital transformation in terms of maturity level. Digital transformation is defined as "the use of technology to radically improve performance or reach of enterprises" (p. 5). Based on interviews with 157 executives in 50 companies, the study defined four level of maturity based on two criteria: digital intensity and transformation management intensity (pp. 60–62):

- *Digital beginners*: firms with low scores on both criteria
- *Digital fashionistas*: firms with high levels of implementation in terms of digital intensity, but low levels of transformation
- *Digital conservatives*: these firms "represent the wise old men and women of the digital world" (p. 61). They are aware of the importance of digital transformation, but work is still clearly divided into silos.
- *Digitari*: firms that really understand the value of digital transformation and how to take advantage of it.

The authors propose two dimensions as the key building blocks for defining digital transformation: digital intensity and transformation management. This taxonomy is helpful in defining the digital transformation at the firm level, and provides an overall picture of executives in terms of maturity. However, these criteria need to be defined in more detail, and the approach must be extended beyond the traditional boundaries of the (large) firm.

We now look in more detail at how practitioners approach the issue of digital transformation in general. In two recent books (one about the digital enterprise and the other about digital culture), CIGREF defined specific aspects of what they called the "key dimensions of digital transformation".

In the book about the digital enterprise (Ménard 2010), the author justifies the use of the term 'digital enterprise' on the basis of the dissemination of digitality into every process and section of the enterprise (p. 40). The digital enterprise is defined in terms its dimensions: innovation, customers, resources, change management, marketing, and distribution.

These topics were expanded upon in the book about digital culture (CIGREF 2013), which defined a clear agenda phrased in terms of organizational design, new incentive systems, and new forms of leadership.

From this, we can see that digital transformation is an issue that needs further conceptual refinement, especially with regard to its nature, scope, and implications for decision making in organizations.

2.3.1   The Transformational Nature of Digitality

How can we analyze the transformational nature of digital technology? Lucas et al. (2013) based their analysis of transformation on three sectors of activity (financial markets, health care, and consumer experience), while Dehning et al. (2003) analyzed the conditions for technology to be transformational, based on a literature review. They identified the following key criteria: a profound change in the traditional way of doing business, the need to acquire new capabilities, and fundamental changes in tasks. The authors develop several different dimensions of transformation, including: processes; new organizations; relationships; user experience; markets; customers; and the disruptive impact. These dimensions can be analyzed at different levels: that of the individual, the firm and the overall society or economy. For example, in the financial services sector various transformations have been proposed, notably in the stock market. One example is the user experience, where the change from the use of telephones to a purely digital platform has created new, e-trading organizations.

This analysis helps to identify various issues that are relevant to managers. In particular, it highlights the importance of widening the perspective used for the analysis of the use of digital artifacts in and around organizations and ecosystems. The issue of digitization goes beyond entrepreneurship in the traditional sense; it has become ecosystemic. This leads to new perspectives and insights that need to be better integrated through research and action; this focus of the ISD program is consistent with the literature.

2.3.2 Digital Transformation: Its Scope, Scale and Sources

Several dimensions of digitization—and therefore of digital transformation—have been proposed. This section expands on the special issue of MIS Quarterly dedicated to "Digital Business Strategy: Toward a Next Generation of Insights" (June 2013).

Digital transformation relates to four aspects of the firm's business strategy (Bharadwaj et al. 2013):

- Its *scope*, which needs to extend beyond the traditional boundaries of the firm (supply chains, industries, etc.);
- Its *scale*: the emergence of platforms that create important network effects in a context of data abundance;
- Its *speed*: the launch of products/services, decision making, building networks, etc.;
- The *source* of value creation and capture (data, networks, digital architecture).

With respect to *scope*, there is a need to develop new approaches to ecosystems. The concept of a business community that goes beyond an orchestrated platform has been proposed and is defined as "a set of possibly overlapping ecosystems in a defined area of business activity" (Markus and Loebbecke 2013: 650). There is also the need to integrate a community dimension into the digital strategy (Lucas et al. 2013). Such an extension is essential, especially when we consider the key characteristics of what Keen and Williams (2013) call 'ultrasuccessful' firms. Their study provides interesting insights into how value is created in the digital world through an examination of extremely successful firms such as Amazon, Expedia,

Table 2.2 Digital transformation: dimensions, issues and implications for managers

Dimension of digital transformation	Questions for managers (strategic, organizing, business models)	Key topics
The scope of digital strategies	• What analytical approaches go beyond the extended firm view? • What are the emerging spaces for value creation?	Defining and analyzing spaces for value creation
The scale of digital strategies	• What is the relative importance of platforms? What typology? Which governance structures foster innovation?	Defining and analyzing the new scope of value creation
The speed of digital strategies	• How to define and deploy innovative offers?	Analyzing acceleration as a systemic phenomenon
The sources of value creation based on digital strategies	• What are sources of value creation in digital spaces?	Defining how value is proposed in digital spaces

Google, and Facebook, which they contrast with 'ultrafades', i.e. firms that were dominant but have lost ground (e.g. Dell, RIM, Nokia). They take as a starting point four aspects of value: (1) the buyer determine value; (2) its relative and shifting nature; (3) companies' leverage of ecocomplexes; and (4) entrepreneurs who continuously offer new dimensions of value. A transitory and ecosystem view of value that integrates social media is important in this respect (Oestreicher-Singer and Zalmanson 2013).

In terms of *scale*, the power of digital markets lies in platforms that compete to leverage the key resource of the intangible economy: data. In this context, *speed* (acceleration) plays a critical role in success. Acceleration calls for the development of a new form of capital: the capital of digital systems is defined as the cumulative knowledge of the design of IT artifacts, which should be distinguished from, for example, discrete strategies used to enlarge the cumulative stock of knowledge (Woodard et al. 2013). Finally, new digital strategies call for the definition of new governance structures (Drnevich and Croson 2013), together with new forms of leadership (Bennis 2013) (Table 2.2).

2.4 Some Insights from Recent Foresight Programs

This book and the ISD program focus on the future design of enterprises (the 2020 enterprise) and their digitality. While the design is based on ad hoc research, it is also worthwhile integrating insights from recent or similar international programs. Here we consider two types of programs: those that develop a macroscopic perspective, and those that (like ISD) focus on the enterprise perspective.

In the following, we examine programs with a societal and organizational focus.

2.4.1 Macro and Innovation Foresights

2.4.1.1 Global Trends 2030, Alternative Worlds

The report published by the National Intelligence Council (2012) in the United States analyzes the drivers for change at the global scale and provides scenarios for evolution. Four 'megatrends' are proposed: (1) individual empowerment, (2) the diffusion of power at the global level, (3) demographic patterns that impact growth in aging countries, and (4) the food, water, and energy nexus, as demand for these resources will grow as the world's population grows.

The report develops six game changers. These include: the impact of new technologies, the governance gap, and the role of the United States. With respect to the impact of new technologies (number 5) the report underlines the importance of the big data era. The processing and storage of data, together with cloud solutions will expand massively, while fears of an Orwellian state may lead citizens to pressure governments in the North to limit the power of big data analytics. Megacities will emerge from nowhere. New manufacturing and automation technologies (e.g. 3D printing) will improve productivity and reduce the need for outsourcing. There will be breakthroughs in technologies related to security and resources that are vital for populations (water, food, energy). Health technologies will continue to expand.

Based on this, four scenarios are developed: (1) *Stalled engine*, "a scenario in which the risk of interstate conflict rises owing to a new 'great game' in Asia" (p. XII). In this scenario, the United States and Europe will turn inward thereby losing power at the global level; (2) *Fusion*, in this scenario the United States, China and Europe agree on an agenda for global collaboration leading to a worldwide effort to deal with global challenges; (3) *Genie out of the bottle*, this is a worse-case scenario where inequalities increase between regions and nations. In this scenario, "the EU single market barely functions [...] Cities in China's coastal zone continue to thrive, but inequalities increase and split the party" (p. xiii); (4) *Non Stated world*, this scenario is dominated by the emerging power of nonstate actors such as nongovernmental organizations, multinational corporations, academic institutions, megacities, and wealthy individuals. These stakeholders are global leaders in addressing issues such as the environment, inequality, poverty, anti-corruption, and peace.

2.4.1.2 OECD (2015): Securing Livelihoods for All

This OECD report (2015) discusses the issue of livelihoods and presents five future scenarios. The report underlines the significant progress that has been made in improving global livelihoods, noting that more than two billion people have emerged from extreme poverty. However, despite this progress there are still major challenges for the future, especially with regard to disparities in revenue and financial fragility. The report argues that the world is experiencing massive demographic shifts, notably with regard to the provision of pensions for ageing populations. Environmental degradation is another major problem. From this perspective, technology is seen a source of innovation that can help to address some of the major issues. The report makes key recommendations at multinational, national and local levels (Aubert and Wermelinger 2015).

The following five scenarios are formulated (pp. 118–130):

Scenario 1: Automated North

The main drivers are automation and growing inequalities. The impacts on livelihoods are growth in unemployment and the automation of low- and high-skilled jobs. Migration is an important issue. Important policy and societal issues emerge, particularly related to wealth redistribution, taxation and skills development.

Scenario 2: Global financial crash

The financial crisis in emerging economies is the main driver in this scenario. The impacts on livelihoods are protectionism related to increases in poverty, inequality and insecurity.

Scenario 3: Drought and joblessness in the South

Here, the main driver is climate change leading to drought. The impact is related to the inability of farmers to make a living and an increase in the number of jobless young people. Policy issues concern migration, social protection and increased R&D related to drought-resistant agriculture.

Scenario 4: Regenerative economies

Here, the main drivers are innovation and policies that promote sustainability. Innovation creates demand for jobs in new products and services, and problems relate to the deployment of unskilled and older workers. Challenges relate to equality via access to education.

Scenario 5: Creative societies

The IT revolution and new societal attitudes are among the main drivers. Impacts relate to joblessness created by technology, reduced reliance on institutions, and a workforce this is increasingly oriented towards developing their own skills portfolio. In this context, new ways to measure progress are needed that are less dependent on GDP.

2.4.1.3 The European Patent Office Foresight of the Future Patenting System by 2025

The report concerns the future of the patent system (European Patent Office 2007) and an assessment of the impact of various factors. The report presents several scenarios, based on an evaluation of five principal factors:

1. *Power*: sources of global power; governments, multinationals, civil society organizations, special interest movements, international bodies and the possible emergence of new sources of power.
2. *Global jungle*: as globalization leads to major changes in investments, capital flow, economies of scale, etc., the major question is "Who will survive, and for how long?"
3. *Rate of change*: this reflects the tension between rapid changes in technology, the short-term nature of the political system and the long-term cycle of institutions such as the intellectual property (IP) system.
4. *Systemic risk*: increasing global interdependence in terms of finance, technology and ideas is a source of systemic risk. Here, the question is, "Where are the tipping points that threaten global interdependence?"
5. *Knowledge paradox*: given the dissemination of knowledge, societies are challenging the relevance of the IP system. Here the question is "Are there cheaper, quicker methods for protecting and exploiting knowledge than the patent system?" (p. 4).

Based on these factors the report develops four scenarios for the evolution of the IP system by 2025, as a function of the dominant source of power:

Market rules. Business is the dominant driver, and it says "Yes" to IP.

Whose game? Geopolitics are the dominant driver, the IP system does not have global legitimacy and there are conflicting rules between national and regional IP systems.

Trees of knowledge. Society is the dominant driver. In this scenario classical monopoly rights have no legitimacy.

Blue Skies. Technology is the dominant driver. In this scenario IP reform restores global legitimacy to the IP system.

2.4.1.4   ESPAS–Global Trends to 2030: Can the EU Meet the Challenges Ahead?

The ESPAS report (2013) analyses global challenges and trends for the years to come. Five global trends are identified: (1) the fact that the human race is becoming older and richer, with an emerging middle class; (2) the shift of power to Asia; (3) the technological revolution, especially digitization that is infiltrating every aspect of life; (4) a scarcity of resources (energy); and (5) the interdependence of countries that is not matched by global governance. The report identifies three revolutions, which will make the world less secure: (1) an economic and technological revolution, which, while it offers huge potential for productivity, at the same time presents challenges in terms of job creation, inequalities, and increasing poverty for the middle classes, including in Europe; (2) a social and democratic revolution, due to a more connected and therefore more demanding and critical population; and (3) a geopolitical revolution, with a shift of power to Asia. The result of this is a less secure world and higher systemic risk. These trends create major challenges for Europe, in terms of restricting the economy, promoting a society focused on change and innovation (including via the digital revolution), and reinforcing the international role of Europe.

2.4.1.5 Innovation Futures in Europe

The focus here is on innovation in Europe in 2025 (Innovation Futures 2011).

Five scenarios are proposed, based on an in-depth analysis of the driving factors and their interactions.

- Scenario 0: "Nothing changes". In this scenario there are no major changes in the European innovation landscape by 2025, while the major players experience a shortage of innovative skills (engineers and scientists). By 2020, Europe will lose its pioneering role in environmental technologies and by 2025 there will be stagnation in all fields, including social life.
- Scenario 1: "Unleashing the creative spirit: Europe's Innovative Societies". In this scenario Europe becomes a major player in innovation by 2025, with societies becoming the source of the development of new products and services, while sustainable business and consumption become the norm. Societal welfare is no longer defined in monetary terms. By 2025, social and environmental aspects are fully integrated.
- Scenario 2: "The Exhausted Giant. Europe Innovation Fatigue". In this scenario Europe experiences "innovation fatigue". Policymakers and entrepreneurs stick to obsolete growth models. Innovation still exists, but only in the business domain and is not embedded into the overall innovation system. By 2020 European competitiveness declines and by 2025 there is a dramatic shortage of innovation.
- Scenario 3: "Locally-Driven Innovation in a Nutshell". In this scenario, innovation moves to cities and regions. Local entrepreneurship drives innovation, but mainly for local urban purposes. By 2025, Europe is "back on track" (p. 38), but at the citizen/city level.
- Scenario 4: "Prometheus Unbound: Innovation for Innovation's Sake". In this scenario innovation is open and ubiquitous throughout the European system, but mainly for economic purposes. Environmental issues are not comprehensively targeted. By 2025, Europe is a leader in innovation, but protests emerge related to the failure to take into account environmental issues.

2.4.2 Digital Foresights

2.4.2.1 The Digital World by 2030

According to work by the European Internet Foundation (2014), by 2030 we will move to a knowledge society where real time is the dominant factor.

The position of Europe with respect to the knowledge society will depend upon its transformational capacity—a move from mass collaboration to a "knowing society". A fully-powered digital economy is expected to be in place by 2030, through a set of specific changes such as 'smart cities'.

2.4.2.2   Internet Foresight by 2030

According to a report edited by Gille and Marchandise (2013), the future of the Internet has more to do with machine-to-machine interaction. This is expected to have a massive impact on all sectors of activity, individuals, and societies. Because of this, public policies must be fundamentally reworked in order to cope with the challenges.

2.4.2.3   The Future of the Internet in 2025?

Based on inputs from an online Delphi survey of experts and analysis, this study by the Oxford Internet Institute (2010) proposes four scenarios:

1. *Smooth Trip*: the rise of the overall Internet economy.
2. *Going Green*: Internet technologies target environmental needs.
3. *Commercial Big Brother*: the Internet becomes a mainly commercial platform.
4. *Power to the People*: the Internet becomes mainly a forum for democracy and the exchange of knowledge.

2.4.3   Digital Enterprises Foresights

Little work has been carried out into organizing for digital firms. The following three studies are related to the ISD program's objectives and content.

2.4.3.1   Inventing the Organizations of the 21st Century

In their report, Malone et al. (2003) refer to the MIT's "initiative on inventing the organizations of the 21st century", which was sponsored by large companies such as BT, EDS/AT Kearny, the National Westminster Bank, the Norwegian Business Consortium, the Union Bank of Switzerland, and Siemens Nixdorf among others. This five year research initiative (1994–1999) analyzed the transformation in ecosystems, especially between large companies and startups, trends in outsourcing and the decentralization of activities. The report is a collection of contributions that address different facets of organizing: the boundaries of the firm and the role of IT in business performance, and scenarios for 2015.

2.4.3.2 The Future Internet Enterprise Systems (FInES)

The FinES Research Roadmap 2025 (FinES 2012) developed a research agenda within the FinES cluster. Four conceptual dimensions were considered to be critical: the socioeconomic space; the enterprise space; the enterprise system, platforms and application space; and the enabling technology space (ICT solutions, in particular those related to the future of the Internet). For each of these dimensions, the document suggests topics of interest for researchers.

2.4.4 A Synthesis

From these foresight programs we can identify areas of uncertainty, and therefore where there are questions about the future. Of these areas, the issue of global equilibrium and the impact of global change on growth is an open question, as is the impact of digital technology on job creation. With respect to Europe in particular, there is also the issue of the dynamics of its innovation systems and its capacity to take advantage of potentialities offered by sustainability technologies, including digital technologies.

At the microeconomic/sectoral level, digital transformation, due to the Internet, is already underway in many sectors, and significant changes are expected, especially with regard to machine-to-machine interaction, while business models are undergoing continuous change in many sectors.

The ubiquity of digital technology is paving the way for new forms of business organization. Therefore, we need to refine their content, structure and building blocks. This will be addressed in the next chapter.

Chapter 3
Key Topics, Emergencies

Here will take the 2020 as a perspective for analyzing digital transformation of firms and organizations. It is based among other things on "weak signals" identified by the ISD international projects. This chapter presents the program's main research topics, before considering the elements of emergencies, especially with regards to the key issues of value creation, organizational design, coordination of work and entrepreneurship. The chapter presents into details these emergencies and put them into perspective.

3.1 The Key Themes of the ISD Program

As with any research program, ISD was built around a series of questionings, translated into the research agenda and updated at each call for projects (projects Wave A, Wave B, Wave C are presented in Annexe A).

3.1.1 Thematic Positioning of Each Project

Figures 3.1, 3.2, 3.3, 3.4 and 3.5 indicate the key themes in each project of the ISD research program. This thematic classification is used for a schematic representation of the program around key words and themes.

The themes are at once *conceptual* (modular approach, acceluction, generative machine, co-creative innovation, improvisation), *functional* (social networks, working conditions, innovation, privacy, standards, governance, copyright), and *geographical* (Asia, Europe).

The themes developed highlight the extreme diversity, as well as the ubiquity, of Digital as a field of study.

© Springer International Publishing Switzerland 2016
A. Bounfour, *Digital Futures, Digital Transformation,*
Progress in IS, DOI 10.1007/978-3-319-23279-9_3

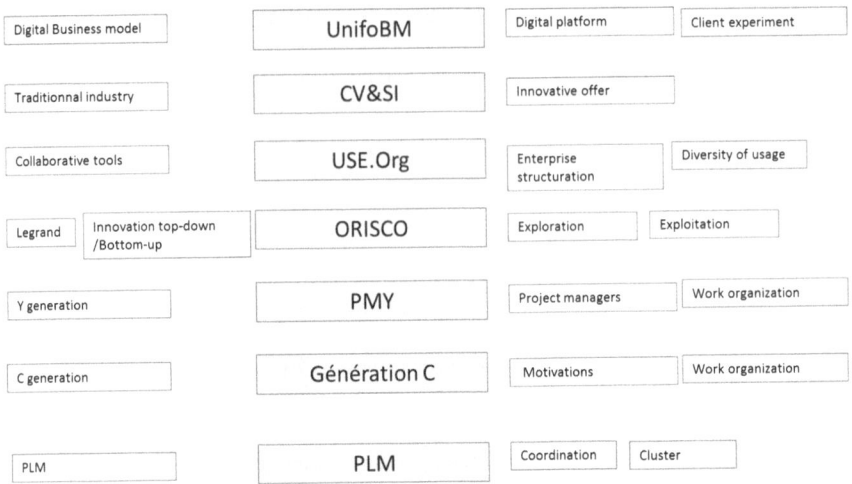

Fig. 3.1 Thematic map of ISD projects

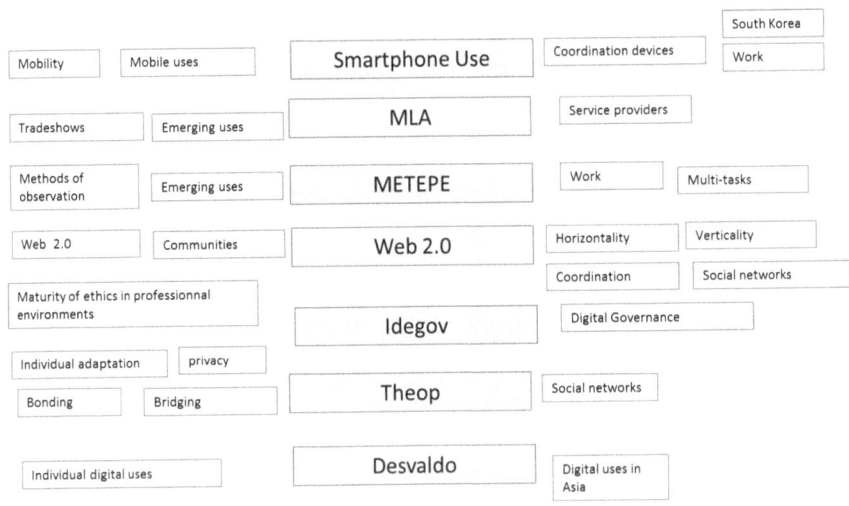

Fig. 3.2 Thematic map of ISD projects

A notable feature of the program is its broad international coverage; the analyses involved concern not only the OECD countries (mainly the USA and Europe), but also the Asia zone, represented here by its main countries: continental China, Taiwan, South Korea, and Japan, both in terms of the scope of analysis and experimentation and in terms of research teams.

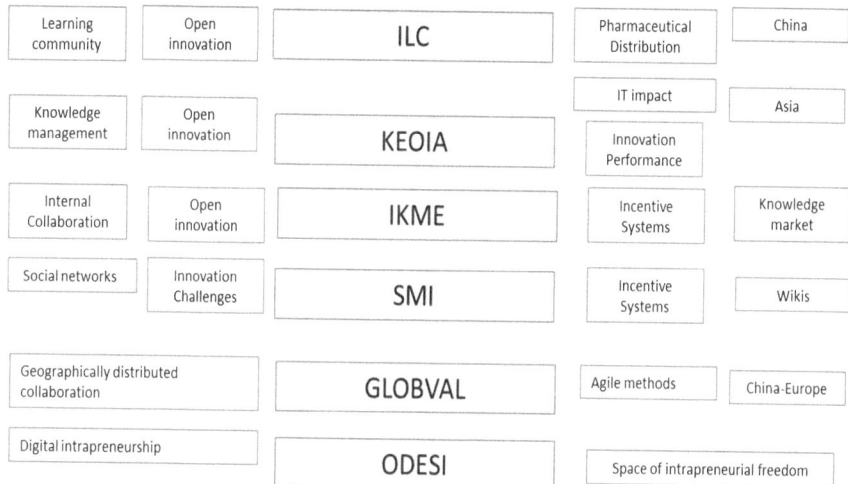

Fig. 3.3 Thematic map of ISD projects

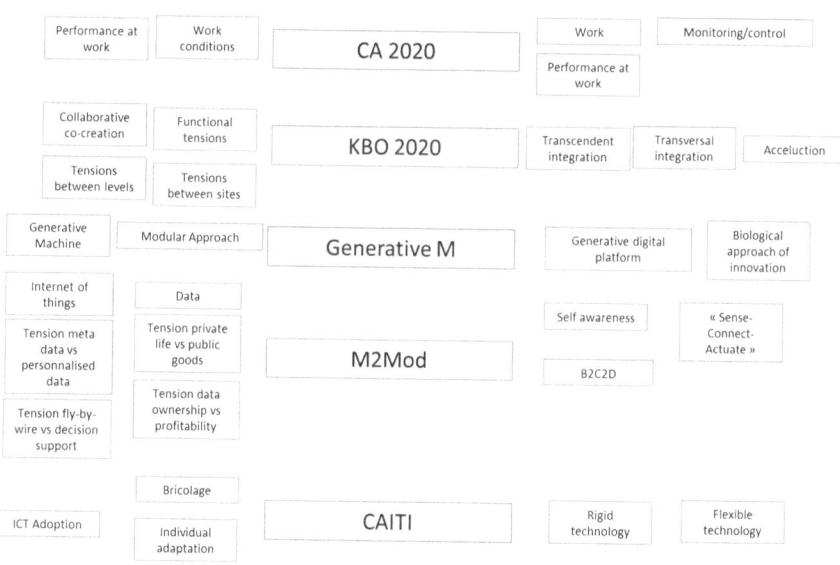

Fig. 3.4 Thematic map of ISD projects

An important characteristic of the ISD program should be underlined here. The analyses developed by the program, and the associated proposals, do not reflect a purely "Western" vision of digital. They also integrate analytical developments and practices from the main economic zone driving the growth of the global economy:

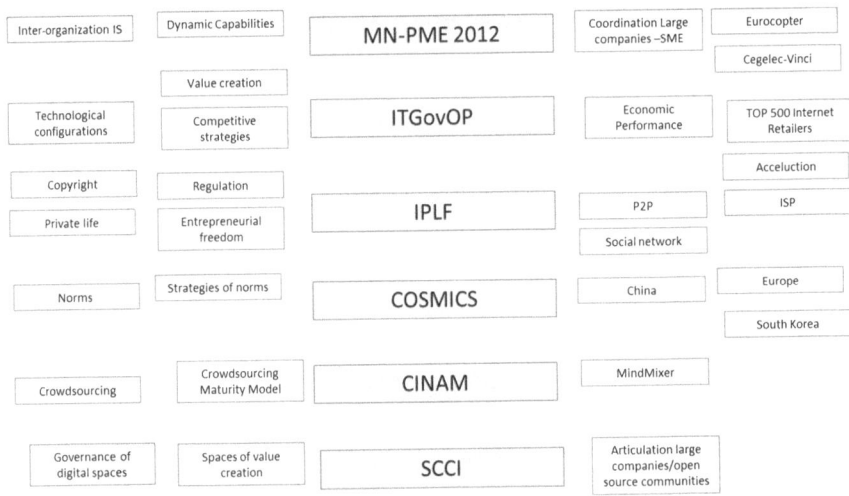

Fig. 3.5 Thematic map of ISD projects

Asia. The propositions developed here are not therefore the exclusive product of a Western vision and practice of digitality; they encompass a wider vision of the geography of uses and related trends.

The details and the acronym of each project are specified in the appendix.

3.1.2   Thematic Clustering

The grouping of themes into clusters serves to indicate the generic themes of the ISD program as well as specific contribution of each project.

Overall, in addition to the breakdown of the program into twelve work packages, the contributions of the thirty or so projects in the program can be structured into eleven generic themes (Fig. 3.6).

3.1.2.1   Business Models and Innovation Ecosystems

Several ISD program projects relate to the theme of business models and the development of innovation ecosystems. A model is proposed (VISOR, UnifoBM) that is complementary to the tools available on the business model design market (notably BM Generation Canvas). Other projects analyze the way the information systems of large companies articulate with SMEs (MN-PME), and the modes of development of cooperative innovation systems around PLM tools (PLM Taiwan).

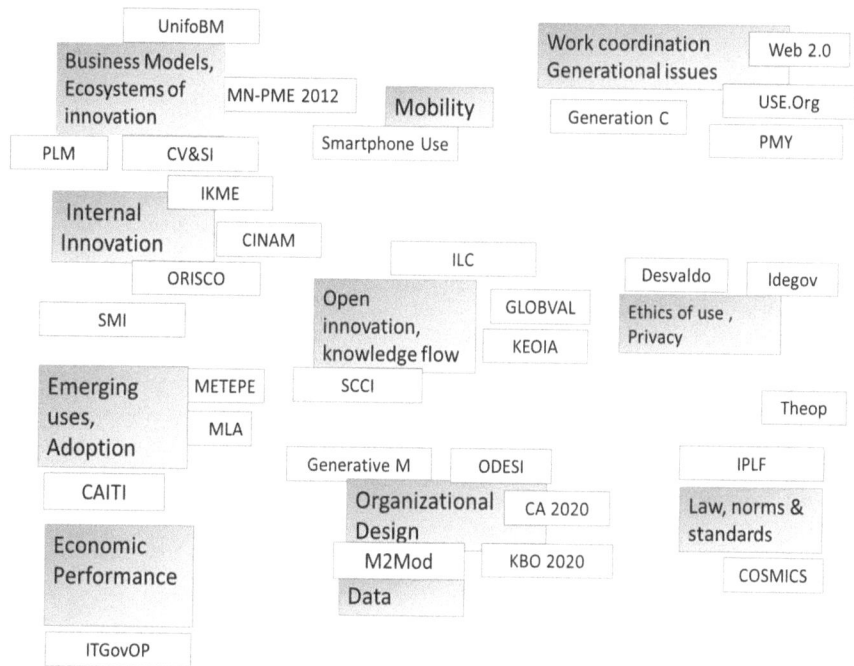

Fig. 3.6 Cluster map of ISD projects

3.1.2.2   Mobility

Mobility is the subject of the SMC project, developed by Namjae Cho and his team, from the University of Hanyang (Seoul, South Korea). Building on the existing literature, and on a wide-ranging survey conducted in South Korea, the project points to the importance of mobility as a critical dimension in intra-organizational coordination. It indicates how the design of mobile uses needs to be defined and deployed around the nature of the tasks involved (and in particular their level of complexity).

3.1.2.3   Work, Coordination and the Generation Question

Work, coordination and the generation question (Generation Y, Generation C), are central topics of reflection and action for businesses. Four of the projects develop analyses on each of these dimensions: the Web 2.0 project evaluates the concept of enterprise 2.0 by spotlighting its benefits and its ambiguities; PMY looks at the behavior of Generation-Y project managers; the "Generation C" study examines the behavior and expectations of Generation C in Quebec; and finally CA2020 conducts and evaluates an experiment in modes of control at a call center, with a relaxation in digital modes of control.

CA2020

Centre d'Etudes de l'Emploi. *Centre d'appel 2020* (Nathalie Greenan, Isabelle Gillet, Rémi Le Gall).

The project involved a life-size simulation of a surveillance (and non-surveillance) system at a call center. The experiment was carried out with the cooperation of 44 volunteer telecounselors at one establishment (80 % of the workforce). Its purpose was to evaluate the effects of organizational change—in the form of surveillance systems—on working conditions. It consisted in reducing the intensity of IT surveillance at the individual level, by modifying the software settings on the tele-counselors' workstations, and on those of their supervisors. The results were subsequently compared with a control group for which no such change had been made.

For the telecounselors, the changes meant that they no longer had access to their individual statistics in real time, or to queuing information.

The supervisors no longer had access to individual statistics in real time, or to information on the duration of each state.

3.1.2.4 Emergent Uses and Individual Adaptation

Three projects come under this heading: METEPE, MLA and CAITI. The first two deal with the modalities of observing emergent uses: whether via purpose-built platforms (METEPE), or by analyzing discourse in traditional spaces; i.e. trade fairs (MLA). CAITI, meanwhile, looks at the central question of how individuals adapt to specific technology profiles: rigid ERP-type technologies versus flexible technologies (such as Google Apps). Individuals are found to adopt *ad hoc* adjustment (and improvisation) behaviors according to the type of technology involved.

3.1.2.5 Internal Innovation

This heading covers projects that seek to analyze how digital-oriented internal innovation is managed within the enterprise. Four projects probe this specific dimension, with different focuses: IKME looks at the enterprise as an internal market for digitally-mediated knowledge creation and development; ORISCO analyzes internal innovation modalities from the perspective of ambidexterity; ILE assesses the conditions of use of social media (wikis) for the deployment of innovation challenges; and KBO 2020 examines the conditions for the development of innovative products and services with short time constraints and geographically distributed resources (teams).

KBO 2020

University of Washington, University of Southern California (Paul Collins, Ann Majchrzak)

Sociotechnical Designs for 2020 R&D Enterprises: Accelerating Innovation by Emergently Leveraging Global Distributed Knowledge, Human Capital, and Digital Assets.

The scope of the project is the R&D activities of a large European electronics group with a strong international presence. Three R&D programs, at differing stages of advancement, were observed over an 18-month period.

Drawing on field observations, the study develops a theory of collaborative co-creation for the R&D intensive enterprise of 2020. These organizations are rendered complex by the need for repeated and continuous management *of alignment tensions*, notably between teams: an exercise made more difficult by disciplinary, functional and geographical diversity. These tensions, and market volatility, make this alignment task extremely challenging. The study concludes, therefore, that *the 2020 enterprise should focus its efforts not on aligning its teams, but on encouraging them to integrate forces both favorable and unfavorable to alignment.*

3.1.2.6   Open Innovation and Knowledge Flows

Open innovation involves the use of resources external to the enterprise to develop innovative products and services. Open innovation can be geographically—even internationally—distributed, raising questions about the management of knowledge flows between entities with different geographies, knowledge levels, or cultures (notably in a Euro-Asian context).

Four projects are grouped together under this theme: ILC looks at the training processes of a learning community in the China's pharmaceutical distribution sector; GlobalVal examines modes of collaboration in IT development projects between Europe (the UK) and China; KEOIA assesses modes of development of open innovation in the Asia zone, and the specific role of information technologies; and SCCI evaluates and develops a model of exchanges between enterprises and open-source communities, and proposes a suitable mode of governance. In a related area, CINAM analyzes the modalities for the use of crowdsourcing to generate new ideas.

M2MMod

University of Southern California (Francis Pereira, Omar El Sawy, Ron Ploof)

Learning from M2M Business Models: Implications for the Business Enterprise 2020. The project proposes a design for the 2020 enterprise based on the prospect of "networked abundance" as defined by five criteria: the three Vs (volume, velocity, variety) of data; connectivity at the edges; autonomic networking (self-awareness) of digital objects; and extreme customization.

It looks at how the abundance of networks requires us to be aware of **four tensions** central to value creation in the 2020 enterprise:

- The *tension between the autonomy of objects ("fly-by-wire") and decision support*;
- The *tension between security and privacy*;
- The *tension between data ownership and profitability*;
- The *tension between public goods* (data as a public good that everyone can use) *and private goods* (the appropriation of data by economic actors, including businesses).

The proposed organizational design leads to a distinction between four types of enterprise:

- Profile 1: *Network abundance enterprises*, which develop a strong position on the two considered criteria (Google being cited as an example);
- Profile 2: *Digital customer orientation enterprises*, which are naturally less focused on the general connectivity between the physical and virtual worlds (e.g. Amazon, FedEx, Volvo);
- Profile 3: *"One-world seamless enterprises"*, which, by contrast, are more focused on exploiting the links between the physical and virtual worlds than on leveraging the customer experience (e.g. Kaiser Permanente, an American health firm);
- Profile 4: *"Two-world separate digital enterprises"*: those enterprises in which the development of digital, especially in its experiential aspects, is of no special benefit in terms of competitive advantage. According to the researchers, between now and 2020, it is unlikely that oil companies such as Occidental Petroleum or Halliburton will adopt the kind of network-abundance model described in the study.

3.1.2.7 The Ethics of Digital Uses and Privacy

Three projects develop this theme, with its predominantly social focus: Desvaldo analyzes public uses of digital objects in the Asian context (China, Taiwan, Japan, South Korea), adding a comparison with the UK; Idegov develops a theoretical approach to the ethics of use, and analyzes its application to the workplace, with an international comparison; and THeop models the ways in which individuals expose their private lives in social networks.

3.1.2.8 Norms, Standards, and the Law

The question of the evolution of the legal context with regard to digital innovation is currently being debated in several institutional and professional contexts, both at the European level and within the OECD framework. In the specific case of digital, the core issue is the adaptation of copyright systems. This legal question leads naturally to the matter of standards, and particularly to the level and mode of participation of enterprises, and to their definition and deployment in the competitive context of digital as we know it.

The IPLF project deals with the question of intellectual property regimes—and of copyright in particular—and their evolution in the international context, while Cosmics analyzes the question of standards and their place on the business agenda, notably in the European context.

3.1.2.9 Economic Performance

The topic of value creation by information technologies is widely analyzed in the literature, but without providing us with a stable analytical framework and established results. ITGovOP models the determinants of the contribution of information technologies to businesses' economic performance, with reference to a specific sector: that of e-commerce in the USA. MN-PME 2012 analyzes the impact of coordination between large corporations and their SME partners, particularly in terms of ecosystemic performance.

3.1.2.10 Data

To say that the data question lies at the heart of the digital transformation of business and of all public and private ecosystems—and indeed of society in general—is a truism. The issue of emergent data ecosystems and their governance is central to the agenda of business leaders. M2Mod analyzes this issue and deduces the implications and potential scenarios for the 2020 enterprise.

3.1.2.11 The Design of the 2020 Enterprise

All of the projects aim to formulate outline proposals for the design of the 2020 enterprise, but some specifically focus on particular elements of its design. ODESI puts forward proposals to develop entrepreneurial space. Generative M analyzes the generative nature of digital and its modes of dissemination and innovation, and proposes a design mode for the 2020 enterprise that integrates this specificity; KBO analyzes the space-time constraints on new product development in the large 2020 enterprise, and what this means in terms of coordination; CA 2020 conducts and evaluates an experiment into a mode of governance of a call center where the digital

control rules were modified (relaxed); and finally, M2MMod looks at the impact of
network abundance on the design of the 2020 enterprise.

GENERATIVE M

Temple University (USA): (Youngjin Yoo, Rob Kulathinal, Sunil Wattal) Designing
21st century Organizations for Generativity: An Organizational Genetics Approach

Generative M's contribution is conceptual as well as methodological. The questions
it poses are: How are digitally-driven innovations to be characterized scientifically?
and What organizational design elements can be proposed? It is on these two points
that the project makes essential contributions, both to the methodology and to the
design of the 2020 enterprise.

In methodological terms, the project adopts a biological approach in order to
understand the mechanisms behind digital innovation. From this analysis, the project
develops a number of elements central to the understanding of the dynamics of
digital innovation, and, naturally, to the design of 21st century enterprises:

- The innovations of the 21st century will be enabled by "generative machines":
  machines capable of producing *spontaneous, unexpected* innovations *often led
  by external developers*. Clearly, we are going to have to revisit the question of
  control in organizations, and the importance of developing the capacity to adapt,
  and to absorb external knowledge and innovations, in a decentralized and
  unplanned way.
- From this starting point, the study formulates recommendations for enterprises, to
  encourage them to prepare for the transition to the generative digital platform model:

  - *Focus on creating basic modules whose components facilitate innovation*
  - *Establish rules to enable these modules to combine or mutate*
  - *Set limits so that the scalability of the modules is linked to the strategic
    objectives of the enterprise* (the objectives of external developers are not
    always aligned with those of the organization)
  - *Use genetic techniques such as sequential analysis and phylogenetics to
    measure and guide development.*

3.2 Digital Emergencies

The ISD projects suggest several key topics related to digital emergencies. Some of
them are particularly relevant to digital futures, and here we propose five.

3.2.1 Innovation and Business Modelling Ecosystems

To an increasing extent, value is created in innovative ecosystems, while thinking
continues to be based on a single unit—the firm. An important intellectual shift is
therefore required, which acknowledges better integration of innovative ecosystems

as the key matrix for decision making. As underlined earlier, there is already a shift in the scope of value creation towards a more platformic form of organizing.

3.2.2 Entrepreneurship

As organizations become increasingly boundaryless, entrepreneurship is both a problem and an opportunity. The multiplicity of interstices within the overall innovation ecosystem opens up such opportunities, which also pose serious problems for contractual arrangements especially within large organizations.

3.2.3 Abundant Data

Data are new sources of growth for companies and organizations. They are one of the main ingredients for future organizing, especially in the context of machine-to-machine interconnection. Data are also an issue from the societal point of view, especially with regard to the way large platformic networks deal with privacy requirements, and the expectations of citizens and stakeholders.

3.2.4 Work in Digital Worlds

Meda (2010) focuses on the changing content of work and working conditions. However, this approach needs to integrate digitality to a greater extent. Working in a context of digital ubiquity is a major issue for the future, as most modern organizations have been built on the concept of the control of sequential time. Work is becoming less sequential and is no longer the province of mono-employees and organizations. In this context, new ways of socio-contracting are called for, while innovation at this level has yet to appear, especially in a context of freelancing and multi-employment.

3.2.5 Regional Specificities

While digitality may be a global phenomenon, its effective deployment obeys specific regional characteristics. We therefore need to understand how Asia (China) is using digital resources and platforms in comparison to Europe, North America and other parts of the world. Developing a regional agenda for digital futures is a major issue in understanding the real economic, geopolitical and cultural stakes.

The next chapter considers in more detail 25 major trends, which have emerged from the program.

Chapter 4
25 Major Trends

The chapter highlights key IT trends for the digital enterprise. These tendencies will be presented along the key topics for enterprises and their management: business modelling, organization of work, ethical and societal use of digital artifacts, technology and data, regulation. 25 tendencies are identified. They are presented and argumented into details, based on the results of projects, but also taking into account the feedback received from Executives and scholars in different arena.

4.1 Transformation Factors: ISD's 25 Propositions

The work of identifying and evaluating emergent digital transformations drew on a corpus of some thirty research projects, conducted under the direction of top-flight international teams. The projects were selected on the basis of a research agenda structured into a dozen work packages and three waves (A, B and C; see appendix for the projects, see Annexe A).

Analysis of the content of the projects throws up a number of points that crystallize emerging practices of great significance for the coming decade. Specifically, these points concern business models, work organization and HR, open innovation practices, ethical issues around private digital uses, knowledge flows, intellectual property rights, and data. There is no need to go through each of these points here—they are developed in the general synthesis reports for the Wave A, Wave B and Wave C projects—but it is worth selecting a few that are of particular significance for the 2020 enterprise.

The following twenty-five propositions are formulated around the key themes of the program.

© Springer International Publishing Switzerland 2016
A. Bounfour, *Digital Futures, Digital Transformation*,
Progress in IS, DOI 10.1007/978-3-319-23279-9_4

4.1.1 Emerging Business Models

The projects sponsored under the ISD program provide grounds for propositions regarding the structuring of digital space in the enterprise of the year 2020:

1. Business models are undergoing a major transformation, without any fundamental difference appearing between purely digital models and traditional models

Both the review of the literature and the case studies analyzed suggest that no clear differentiation can yet be made between traditional model and emergent models; more precisely, one finds that traditional models nonetheless contain elements of emergence. The five cases studied by the program clearly show an ongoing transformation of business models by digital, notably with a collaborative dimension and with an increasingly evident integration of horizontal, vertical or oblique business lines. The existence of a duality of adoption in certain enterprises (for example, in the oil industry) is underlined.

2. The expansion of value creation spaces is a major trend driven by digital

This expansion is essential: it depends on adaptive business models. Value, meanwhile, can be created in multiple ways, combining transactional, but also relational, resources—of which the dissemination of information and knowledge is a key component. Expansion is also made possible by the emergence of new collaborative tools such as PLM, and made necessary by the need to integrate data and information systems between clients (large companies) and their suppliers (SMEs in particular). Trends such as these call for us to think in terms of "value creation ecosystems" rather than in terms of value chains.

3. The "customer experience", mediated by digital platforms, is an essential dimension of emerging digital uses

As the case of Nike in the USA (along with other business examples) attests, the customer experience is a key component of emerging economic models. Hence the attention paid to data in effective digital models, but also to platforms centering on the customer experience.

4. "Pure" digital business lines (software and services) appear to be quicker to develop adaptive offerings than more "traditional" business lines. New business models call for new managerial skills

Changes in business models, continuous innovation in products and services, and the involvement of differentiated value creation spaces naturally call for appropriate systems of management and leadership. These include, for example, managing new forms of customer relationship, with multiple relational and professional identities, managing industrial property rights, and managing the tension between the company's "internal" resources and "external" resources that are (or can be) mobilized for cooperation or for spot market transactions.

4.1.2   Work, Coordination and Digital Uses

When it comes to organizing the 2020 enterprise, the question of work organization and the coordination of the related tasks are plainly essential. The program suggests four propositions in this respect:

5. In enterprise 2.0, hierarchy and horizontality (community) are not naturally opposed: they are two distinct but complementary modes of coordination

Analysis of the literature and evaluation of business case studies calls into question some of the received wisdom, notably the idea that hierarchy and community are in opposition (in fact they relate to different coordination mechanisms). Likewise the idea that communities can be seen as the materialization of enterprise 2.0 (communities have an identity of their own).

6. Digital technology is relatively neutral with regard to collaborative uses; uses are also (above all?) driven by organizational specificities. Individuals adapt to the nature of technology by improvisation, especially where the technology is flexible

Types of digital uses can depend on professional-identity situations (for example, in an aeronautical firm that was studied, the secretaries have developed a collaborative approach, while the research personnel have not, for reasons connected to the nature of the proposed tool). Similarly, decentralization is not necessarily a determining factor, due to the ambiguity of the technology: it can lead to the implementation of standardization in decentralized structures or to "slack" (the "wiggle room" that players create for themselves outside of formal rules) in large companies with a centralized (bureaucratic and matricial) or divisional structure—or it can reinforce existing modes of organization. At the individual level, people adapt to uses by "improvisation", in accordance with rules that are not always determined in advance. This is particularly true of flexible technologies (Google apps, in this case).

7. Mobility: the revolution is already visible, but its transformational impacts are still taking shape

Mobile uses have massively penetrated the professional sphere, and are now one of its structural components. Mobility is becoming a promising new space for value creation. Compared to "fixed" workforces, giving employees mobility multiplies their productivity, while wrenching them away from administrative tasks. This emerges clearly from research conducted in Asia (notably in South Korea).

8. The behaviors of the new generations (Y and C) reflect new digital uses… but they also derive from earlier managerial practices

The generational question is critical to understanding the dynamics of digital uses in work-related spaces. The research conducted for the program indicates certain heterogeneity of profiles among Generation Y when it comes to networking practices, the relationship to time, and project management. They also suggest a relationship to the enterprise based on "something-for-something", reflecting a distantiated relationship with the enterprise, tied in with memories of past managerial practices experienced by Generation Y's parents (restructuring, outsourcing, etc.). Generation C, by contrast, seems more concerned with the question of time management, the control of time by the enterprise, and thus of the division between personal time and working time.

9. The impact of digital uses on the question of control needs to be approached indirectly, taking due account of the nature of tasks

For Taylorized tasks, a relaxation of control can be destabilizing for the people concerned (as in the case of call centers). For complex tasks, on the other hand, control can be understood as meaning greater autonomy for actors. In many respects, control has changed shape: it can be digital for routine tasks (e.g. call centers), but centered on listening, empathy, and leadership, for more complex tasks.

4.1.3 Internal Innovation Practices

10. Innovation is a complex process, in which *top-down* approaches intermesh with *bottom-up* approaches

The ISD research highlights the importance for companies to deal with the double constraint of exploring new paths to innovation while at the same time exploiting their existing capabilities. In one case study, two currents of innovation are observed, one "bottom-up", the other "top-down", with a demarcation of roles between operatives, middle management and top management.

11. Internal collaboration on innovation can be far from straightforward, especially in contexts dominated by individual incentives

The impact of the spread and deployment of web tools (such as blogs, wikis and social networks) on productivity improvement, innovation and collaboration—especially in the management of virtual teams—is not yet a mature and widespread phenomenon. The main use of social media in the enterprise is currently knowledge-sharing rather than innovation.

The regular use of wikis and other online tools for the co-creation of ideas and for innovative collaboration is still in the future. One reason for the slow development of social media usage inside the company lies in intra-organizational incentive mechanisms, which are too centered on individuals rather than on groups.

12. Digital develops a view of the enterprise as a knowledge market, but it also requires the development of approaches centered on more organic relationships

In terms of organizational forms, we usually distinguish between the enterprise, the market, and intermediary forms (networks, communities). But in the ongoing process of deconstruction of socioeconomic systems, it makes particular sense to explore the enterprise as a market, especially in a context where transactional exchanges are developing around knowledge and its related assets. A knowledge market is defined as "*an IT platform for matching information seekers to information sources, with material and social incentives to encourage efficient exchange*" (Benbya 2013). The analysis highlights the phasing of incentive systems, particularly as regards the preeminence of individual material rewards as compared to collective and community-based aspects. In the exchange processes, the latter are observed to predominate over the former in the simulations. What we have here is an articulation of modes of organization (market versus community) driven by the digital platforms in use. The simulations also reveal the role of anonymous behavior in interpersonal interaction. For the 2020 enterprise, the internal market dimension must be given serious consideration, as the 2020 enterprise is an open organization *par excellence*. However, incentive systems must be designed with care, and financial incentives should be prioritized only for the short term.

13. The development of innovative digital products and services in a geographically distributed context calls for continuous and repeated management of alignment tensions (between teams, disciplines and functions)

Space-time constraints lead enterprises—and R&D intensives enterprises in particular—to develop collaborative co-creation processes, with an emphasis on the teams' integration of factors that are favorable or unfavorable to alignment constraints (whether functional, spatial or skills-related). The 2020 enterprise is an enterprise centered on co-creation, through the ***management of multiple tensions***.

4.1.4 Open (External) Innovation Practices

The new paradigm of open innovation challenges the older, closed-innovation models. Open innovation is based on an innovative ecosystem which relies on self-organization and communication. The dynamics of innovative ecosystems depend on cultural factors such as social capital, flexibility, and attitudes to entrepreneurship and risk-taking; in other words, an innovation culture.

14. Open innovation practices contribute to performance. They are heavily reliant on—and greatly facilitated by—companies' digital infrastructures

IT capacities have a major impact on diversity and on knowledge collaboration, which determine the level of development of open collaboration. This in turn determines business performance. In the Asian context, specifically, open innovation practices are differentiated according to each country's level of development.

15. Joint collaborative platforms emerge in competitive industries by means of ad hoc learning processes

This is particularly well illustrated by the ILC project on China's pharmaceutical industry. The study describes the cooperative dynamic that has developed in the Chinese pharmaceutical sector around a shared digital infrastructure, mainly at the initiative of the government, which sought to use the digital infrastructure to circumvent the opaque (and thus corruption-prone) relationship between distributors and hospitals. The authorities therefore promoted an e-commerce approach, but this proved unable to integrate the diverse but cooperative needs of the operators. In particular, pre-existing social practices conditioned (facilitated) migration to the collaborative digital platforms, which were necessary to meet the specific needs expressed.

16. Online collaboration to develop knowledge can be organized even between people with no previous contact

The development of digital platforms with a collaborative function raises fresh questions about interpersonal and intergroup interaction on these platforms, about the incentive systems that are or might be used, and more generally about the overall governance of interactions between contributors with no pre-established—notably face-to-face—links. The research conducted under the program shows that this can be achieved, by using ad hoc incentive systems.

17. Spot markets, crowdsourcing, communities and hierarchy are complementary spaces, and their overall governance remains to be defined

The design of the 2020 enterprise will articulate three organizational forms: the spot market, communities (which may be more or less organic), and hierarchy, the traditional organizational form of the enterprise, especially for large companies. These three spaces will be articulated dynamically, with the possibility of shuttling between forms, even within the enterprise. Crowdsourcing, meanwhile, is developing as a way of deploying spot markets for ideas.

4.1.5   Enterprise Space and Knowledge Flows

The issues of globalization, outsourcing and offshoring are high on the agenda of top managers, and of CIOs in particular. And yet, the understanding of the dynamics of knowledge exchange in distant geographical and cultural contexts has not been systematically evaluated. The same is true of the enterprise space in information systems in and around companies.

For this specific dimension, the research conducted under the program gives rise to two propositions.

18. In digital space-time, remote collaboration calls for skills-sharing to be redesigned. Ad hoc (agile) design methods make this possible

The transformation of the space-time of value creation is a key dimension of digital, and one that needs to be viewed at the global level. The practical questions that arise relate to the coordination of activities in an international context, the emergence of new players, and the management of space-time in the enterprise. The ISD research reveals, through the example of a Sino-European cooperation program, that spatial and temporal constraints on collaboration can be managed by intensive use of collaborative technologies. This is further facilitated by instilling a shared and balanced vision of contributions, reinforcing the related trust, and by understanding the dynamics of "born multinationals".

19. In the digital economy, IT entrepreneurship requires the creation of ad hoc spaces of freedom, especially by large companies

With the development of open production ecosystems, the question of entrepreneurship becomes a core theme, one with legal, strategic, economic and organizational implications. The research conducted under the ISD program provides a structured analytical framework for digital entrepreneurship, particularly around the concepts of *IT entrepreneurial realization* and *spaces of intrapreneurial freedom.*

4.1.6   The Social and Ethical Dimensions of Use

From the point of view of the ISD program, the question of social and ethical values is important because of the displacement of innovation and change from within the company to society as a whole, as well as the strictly societal issues surrounding usage (ethics, privacy, democracy, etc.). For this reason, research on the social dimension of information system use is of critical importance, including from a managerial viewpoint. It should focus particularly on emerging rules and practices, which may be indicative of new ethical values that could have a strong impact on decision-making.

Three propositions stem from the ISD research in this area:

20. For work-related uses, the ethical issues are still emergent. Users are aware of the importance of the topic, but "double-loop" learning is not yet fully in place

The ISD studies look at the implementation of solutions such as codes of conduct, technical mechanisms, and training. But these solutions have not yet reached the required level of maturity. Specifically, there is, as yet, no questioning of the basic motivations and internal logic of behaviors (i.e. double-loop learning).

21. While mobile uses and ubiquitous connectivity may pose ethical problems for public users, this is far from being a universal issue, especially in Asia

The ISD research demonstrates the widespread use of mobile objects, but also indicates that public users in Asia are more concerned with technical constraints on uses than by ethical questions as such. In general, they accept that the decision on whether to save website data should be left to the operators.

22. The hypothesis of the end of privacy, by a general over-exposure of individuals, is not confirmed. Individuals adjust their exposure behavior to the policies of operators like Facebook, and indeed to the structure of their own social networks

How do individuals behave with regard to the exposure of their personal data, and how, where applicable, do they adjust their behavior to the policies implemented by the main social networks? The generally accepted hypothesis is that the widespread over-exposure of individuals leads to the end of privacy. By analyzing specific characteristics and hypotheses, and data from simulations, the ISD research indicates that the network architecture is a key criterion and that the exposure of personal data is not just about isolated behaviors; it is also about networks. Likewise with the "end of privacy" hypothesis: more connections make people more sensitive to the issue of privacy. Contrary to the dominant theory, individuals adjust their level of exposure to the network structures and policies implemented by the operators.

4.1.7   Data, Intellectual Property, and the Specificity of Digital

Questions of data and intellectual property are central to the agenda of enterprises, public policy-makers and society in general. Data raises issues of law, security, privacy and use, from the viewpoint of value creation. Moreover, the mobility of data can usefully be tied in with questions about the specific nature of digital as a production system radically different to previously identified systems. After examining these dimensions, the program formulates three propositions.

23. Digital induces tensions between regulation and freedom, and between privacy and the freedom to do business. Copyright, in particular, is endangered by the development of open-source and open-access practices

Copyright is challenged by several movements—notably "free and open source", "open access" (OA) and "access to knowledge" (A2K)—that seek to deregulate it and thus restrict its scope of application. The protection of privacy is likewise threatened by the development of surveillance regulations, as well as by operators' practices. The spread of Cloud applications, social networks, and the generalized use of smartphones tends to incite users to "sacrifice" some of the data relating to their privacy. There is a tension between regulation and freedom, especially the freedom of expression and the freedom to do business. This tension is problematic for businesses, notably for Internet service providers, who have to make trade-offs between respect for privacy and the development of data-driven business activities.

24. Digital is a generative machine that produces spontaneous, unexpected innovations, often contributed by external developers

The innovations of the 21st century will be enabled by "generative machines": machines capable of producing spontaneous, unexpected innovations often led by developers from outside the company or innovation platform. From this starting-point, the concept of generative digital platform is proposed as a model for the structure of digital innovation for the 2020 enterprise. The model refers to an organizational structure in which innovation is led by external actors, and the pace of change is continuous. Organizations such as WordPress correspond to this model, with thousands of developers working independently to improve the platform's functionalities.

25. Network abundance multiplies the tensions involved in organizing the enterprise (fly-by-wire vs. decision support; security vs. privacy; ownership vs. profitability; public goods vs. private goods)

The abundance of networks goes beyond the framework of the Internet of Things, as it includes networks of people. The abundance of networks raises four tensions that are central to creation of value by the 2020 enterprise: the tension between the autonomy of objects ("fly-by-wire") and decision support; the tension between security and privacy; the tension between data ownership and profitability; the tension between public goods (data as a public good that everyone can use) and private goods (the appropriation of data by economic actors, including businesses).

The digital choices of the 2020 enterprise will therefore be shaped by the need to manage these tensions. Business profiles will be partly determined by their level of "network-intensity" and thus by their exploitation of links to different spaces.

Building on these propositions, let us now look at the question of the conceptual foundations of digital transformation, a prerequisite to the design of the 2020 enterprise.

II-Key Trends and Executives feedback

These tendencies have been presented and discussed with the governance bodies of the program, especially the strategic committee and the Cigref Foundation board. They also have been discussed in ad hoc workshop as well as internationally.[1] These tendencies have been well received from all the different audiences. The feedback received from Executives, policy makers and scholars, allowed to check the relevance of the general conceptual framework as well as it implications for Executives. Let us consider the work within Eurocio, as an illustration.

During Eurocio 2013 (Brussels), an ad hoc workshop has been dedicated to the design of the 2020 enterprise. Participating CIOs were invited to jointly work along the following questionings:

---

[1]At the international level, these trends and related conceptual levels, have been presented in different arena. Among these: ICIS, Shanghai, 2011, IC8—The 8th edition of the world conference on intellectual capital, The World Bank, Paris, 31st of May 2012; Global Forum, 12 and 13 November 2012, Stockholm; Eurocio 28 and 29 November 2013, Brussels, The CIGREF—DG Connect meeting summit in Brussels, September 2013.

- Which types of value spaces for the 2020 enterprise?
- What types of scenarios to be considered?
- How concretely the next generation of organizational models will impact the IS governance?
- How digital resources will contribute/lead such a transformation?
- What are the implications for profiling the next IS function?

Based on these elements, the workshop discussed the following topics:

- Emerging business models
- Emerging societal values and norms
- Outdoor/open innovation
- Collaborative spaces
- Work, coordination, Web 2.0

The main conclusions of the workshop can be summarized as follows:

- New business models are driven by data, customer experience and new CRM capabilities
- Firm's frontiers are disappearing
- Ethics and privacy issues may be more prominent after 2020
- Open innovation is more and more used

The group converges on most of these tendencies especially with regards to the multiplicity of spaces, the acceleration of links and the comprehensive nature of the overall mapping provided by ISD.

Chapter 5
The Emerging Production System

This chapter proceeds to the thematic analyses of the 25 tendencies, with the objective of identifying the key building blocks of the emerging production system in digital spaces. Such a characterization is necessary to the design of future scenarios of the digital enterprise and its related ecosystems. It then identifies the key building blocks for the new production system that underline the digital transformation of firms and organizations. Five building blocks are identified for the new production system: the multiplicity of value creation spaces, the acceleration of links, the contraction of space and time in the coordination of activities and the preeminence of society vis-à-vis the organization. These five building blocks will be presented and argumented in an articulated way.

5.1 Thematic Analysis of the Propositions

The twenty five propositions arising from the research program, and formulated on the basis of the projects, bring us back to our initial questions: How should the 2020 enterprise be designed? How will its digital resources fit together, if they do, and with what implications? And finally, how will they be governed?

The purpose of ISD, of course, is to design the 2020 enterprise: *the focus is therefore primarily organizational*.

From the propositions presented above, and with regard to the program's basic objective, the time has come to outline the conceptual building blocks of this design. These will be based on an analysis of the detailed content of each project and on the project overviews, as presented in the program's three ad hoc reports.[1]

A cross-reading of the propositions identifies five dimensions that make up the conceptual model:

1. *Value creation spaces*, examining their specificities, their multiplicity and their instability;

---

[1]Bounfour (2011, 2013, 2014).

© Springer International Publishing Switzerland 2016
A. Bounfour, *Digital Futures, Digital Transformation*,
Progress in IS, DOI 10.1007/978-3-319-23279-9_5

2. *Incentive systems*: the explicit or implicit rules (including cultural) required to keep exchanges flowing between and within these spaces;
3. *The space-time dimension of performance*: how do the time and space of production intermesh? How does the acceleration at work in digital spaces influence emerging modes of production?
   The **acceleration of links** emerges here as a key element in the organizational dynamic;
4. *The modes of articulation between the "enterprise production space (or the entrepreneurial space)", and the "social production space"*, or the equivalence of norms: this point is essential, due to the preeminence of "social" spaces in the processes that produce tangible and intangible goods and services in post-industrial societies. **The plasticity** (or liquidity) of organizational spaces proves fundamental here;
5. *The ethics of digital use*, in other words the way enterprises and society in general address the question of digital use, especially on the central issue of data, but also with regard to control.

The program propositions will be contextualized here relative to the importance of the ethics of use in the design of the 2020 enterprise.

Table 5.1 lists in detail the thematic elements to be considered in the conceptual building blocks for the design of the 2020 enterprise. Let us go back over each point in turn.

5.2   An Expansion of Value Production Spaces

The ISD projects have revealed, directly or indirectly, a wide expansion in businesses' domain of value production: no longer limited to the traditional boundaries of the enterprise, it now encompasses the domain of its competitors, additional resources (such as those of suppliers or relatively unconnected sectors), customers, mobility as a new space of production, data, social spaces… and the private time of its employees. Such an expansion suggests, on closer inspection, **a new mode of production, of which digital is a key component, and whose boundaries and guiding principles are yet to be determined**.

Two principles can be announced here as constituting this new mode of production:

* **The expansion of the domain of value production** to multiple spaces, whose boundaries and guiding principles need to be more clearly identified;
* **The instantaneity of exchanges** (transactional or, to varying degrees, organic), driven by the ongoing acceleration of digital.

The consequence of this dual trend is a fusion of the space-time of individual and collective action. Digital uses considerably reduce the space of action through the instantaneity of the space of flows (Castells 2000).

Table 5.1 Designing the 2020 enterprise: the conceptual building blocks

25 trends (proposals)	Underlying organizational design themes			
	Value creation spaces	Incentive systems	Space-time dimension of performance	Articulation of "production" spaces (social vs. enterprise)
Emergent business models				
1. Business models are undergoing a major transformation, without any fundamental difference appearing between purely digital models and traditional models	Multiplicity of spaces	Customer experience	Acceleration of links	Plasticity
2. The expansion of value creation spaces is a major trend driven by digital				
3. The "customer experience", mediated by digital platforms, is an essential dimension of emerging digital uses				
4. "Pure" digital business lines (software and services) appear to be quicker to develop adaptive offerings than more "traditional" business lines. New business models call for new managerial skills				

(continued)

Table 5.1 (continued)

25 trends (proposals)	Underlying organizational design themes			
	Value creation spaces	Incentive systems	Space-time dimension of performance	Articulation of "production" spaces (social vs. enterprise)
Work, coordination and digital uses				
5. In enterprise 2.0, hierarchy and horizontality (community) are not naturally opposed: they are two distinct but complementary modes of coordination	Multiplicity of spaces Mobility: a new value creation space	Transaction/organicity	Acceleration of links	Plasticity Contingent plasticity
6. Digital technology is relatively neutral with regard to collaborative uses; uses are also (above all?) driven by organizational specificities. Individuals adapt to the nature of technology by improvisation, especially where the technology is flexible				
7. Mobility: the revolution is already visible, but its transformational impacts are still taking shape				
8. The behaviors of the new generations (Y and C) reflect new digital uses… but they also derive from earlier managerial practices				
9. The impact of digital uses on the question of control needs to be approached indirectly, taking due account of the nature of tasks				

(continued)

Table 5.1 (continued)

25 trends (proposals)	Underlying organizational design themes			
	Value creation spaces	Incentive systems	Space-time dimension of performance	Articulation of "production" spaces (social vs. enterprise)
Internal innovation practices				
10. Innovation is a complex process, in which top-down approaches intermesh with bottom-up approaches	Enterprise space, geographical space Multiple spaces	Transaction versus organicity	Acceleration of links	Enterprise space
11. Internal collaboration on innovation can be far from straightforward, especially in contexts dominated by individual incentives				
12. Digital develops a view of the enterprise as a knowledge market, but it also requires the development of approaches centered on more organic relationships				
13. The development of innovative digital products and services in a geographically distributed context calls for continuous and repeated management of alignment tensions (between teams, disciplines and functions)				

(continued)

Table 5.1 (continued)

25 trends (proposals)	Underlying organizational design themes			
	Value creation spaces	Incentive systems	Space-time dimension of performance	Articulation of "production" spaces (social vs. enterprise)
Open (external) innovation practices				
14. Open innovation practices contribute to performance. They are heavily reliant on—and greatly facilitated by—companies' digital infrastructures	Knowledge market space, internal and external	Transaction versus Organicity	Acceleration of links	Articulation between internal and external enterprise space and societal space
15. Joint collaborative platforms emerge in competitive industries by means of ad hoc learning processes	Multiple spaces			
16. Online collaboration to develop knowledge can be organized even between people with no previous contact				
17. Spot markets, crowdsourcing, communities and hierarchy are complementary spaces, and their overall governance remains to be defined				
Enterprise space and knowledge flows				
18. In digital space-time, remote collaboration calls for skills-sharing to be redesigned. Ad hoc (agile) design methods make this possible	Multiple geographically distant spaces, entrepreneurship: a new space	Transaction and organicity Autonomous spaces	Spatial contraction	Enterprise spaces
19. In the digital economy, IT entrepreneurship requires the creation of ad hoc spaces of freedom, especially by large companies				

(continued)

Table 5.1 (continued)

25 trends (proposals)	Underlying organizational design themes			
	Value creation spaces	Incentive systems	Space-time dimension of performance	Articulation of "production" spaces (social vs. enterprise)
Social and ethical dimensions of use				
20. For work-related uses, ethical issues are still emergent. Users are aware of the importance of the topic, but "double-loop" learning is not yet fully in place	Multiplicity of spaces	Single-loop learning Technical performance of exchange spaces	Acceleration of links	Articulation with and within social space
21. While mobile uses and ubiquitous connectivity may pose ethical problems for public users, this is far from being a universal issue, especially in Asia		Adaptive behavior of individuals		
22. The hypothesis of the end of privacy, by a general over-exposure of individuals, is not confirmed. Individuals adjust their exposure behavior to the policies of operators like Facebook, and indeed to the structure of their own social networks				
Data, intellectual property, and the specificity of digital				
23. Digital induces tensions between regulation and freedom, and between privacy and the freedom to do business. Copyright, in particular, is endangered by the development of open-source/open-access practices and the multiplicity of legal regimes	Multiplicity of spaces	Management of multiple tensions spaces	Acceleration of links	Articulation between social and enterprise space/network spaces/platform space/data space
24. Digital is a generative machine that produces spontaneous, unexpected innovations, often contributed by external developers				
25. Network abundance multiplies the tensions involved in organizing the enterprise (fly-by-wire vs. decision support; security vs. privacy; ownership vs. profitability; public goods vs. private goods)				

These two elements tie in with several fundamental analytical arguments that further underline the importance of the spatial dimension—and thus of digital spaces—in the new configurations of value creation. The first, obvious, reason has already been stated: as soon as we touch on the use of digital objects and related systems, we need to go beyond the traditional boundaries of the enterprise. Several major innovations have taken place outside the frontiers of the firm, as the omni-presence of social media witnesses. Secondly, digitality engenders a profound and yet astonishingly silent revolution in the way activities are organized and linked together in different spaces, thus challenging the still-dominant analytical approach to the enterprise. The generativity of digital technology is now seen as an important viewpoint, and perhaps even as a substitute for that approach. Thirdly, and more fundamentally, the use of digital spaces offers an immense potential for the creation and extraction of value by enterprises in the market economy, due to the intrinsic nature of digital: (1) it creates a new support for value creation (social media, mobility, data, etc.); (2) it establishes links between existing physical spaces and new spaces and (3) it accelerates the links between the different spaces. These three basic arguments support the idea that digital spaces extend beyond the collabora-tions that take place inside and around organizations: they represent profound transformations not just in business ecosystems, but more generally in our daily lives.

What we are witnessing is the emergence of a new topography of value creation, with multiple spaces that are governed by more or less flexible (and more or less visible) relations of exchange and control.

This multiplicity of spaces is naturally of critical importance for companies' digital strategies, especially for the larger firms, which—until now—controlled most of their own productive spaces.

5.3   The Space-Time Dimension

As Krugman underlined it (Krugman 1988), space is the final frontier of economic analysis. It is a dimension than has been put firmly back at the center by digital. More specifically, the articulation between space and time is becoming a core element in the emerging mode of production, mainly on account of the digital revolution. Analysis of the project propositions indicates that one of the underlying dimensions of the 2020 enterprise is that of the acceleration of links between different spaces, and thus, of a contraction in space-time.

Acceleration was the subject of a post-doctoral thesis by the German philosopher Hartmut Rosa Rose (2010). I echo some of his arguments here, extending them to the digital domain. Rosa distinguishes between three periods of production: pre-modern (before the industrial revolution), modern (corresponding roughly to 20th-century industrialization) and postmodern (late 20th, early 21st century).

These three periods of human development (at least in the West) are charac-terized by rules of social relations, including for the management of productive time. The modern period, in this respect, is characterized by a clear differentiation between work time and free time, with particular attention paid to productive time. The postmodern period is characterized by the emergence of a fusion of times (productive and personal) and, in the process, by the diminishing relevance of time "controls" (as is clear from the study of Generation C in Quebec). This is made possible only by digital acceleration: its instruments (digital artifacts), by their ubiquity and instantaneous, always-on connectivity, induce a considerable expansion of the production space, making attempts to "control" it anachronistic.

5.3.1 Time and Space in Digital Worlds

Territoriality has been defined in terms of control (Sack 1986). Digitality fundamentally challenges such a concept, or at least invites to the consideration of its main components and the way they are articulated jointly or sequently. For organizations (firms), things our now obvious in each of their daily life. But things are also observable at the more macroscopic level. Territoriality and space is also a question of time, e.g. of relationships between spaces. The ubiquity of digitality indeed has an important impact of the space-time relationship in the new production mode. As Le Goff (1977), demonstrated, the organization of time and space in the medieval age was initiated by merchant around the clock time. The productive space-time was aligned along the merchant considerations, and not along the church ones. For a matter of value creation—and therefore of capital circulation—the time was organized along technological instruments (the clocks); (Harvey 1989: 170–171). Harvey (2005) building on Lefebvre work on space (Lefebvre 2000), distinguished between absolute space (e.g. walls, bridges, etc. for the material space), relative space (e.g.: the circulation, the time-space compression) and relational space (visions, emotions). These three distinctions are now challenged by the emergence of digitality. This leads Hassan (2003) to develop this concept of chronoscopic society, according to which there is a conjunction of a neoliberal globalization with information technology to impact the way knowledge is disseminated in advanced societies. This nexus leads to an "information ecology" which impacts individuals, culture and society.

According to this author, we are rapidly moving from a chronological temporality of clock time to digitality compressed real time, what the author, after Paul Virilio, named the chronoscopic time (Hassan 2003: 5). It is important to refer to time and how it impacts our conditions of work and life. The clock times constitutes the central point of reference for our daily life. Work time is the point of reference for individuals, organizations and societies.

In many respects the clock is the central arm of the industrial revolution, according to Mumford, more than the railway or the steam engine (Mumford 1934/1967, quoted by Hassan: 13). ICT revolution compresses both space and time.

The application of the principles of scientific management by Taylor is illustrative of the importance of the clock time in modern societies: workers perform under the control of clocks, they even need to beat the clocks … to survive.

Time is missing for every daily act of people "Being increasingly suspended in the real-time of chronoscopic temporality means a lack of time to read, to study, to reflect, to consider, to concentrate, to debate and discuss, to care, to empathize, to analyze, to interpret, to scrutinize and to sympathize - and more" (Hassan p. 133). In post-modern societies, time is the scare resource: we are missing of time as it has been underlined by Eriksen (2001), with his "tyranny of the moment". The acceleration of everything induces the change in the scarcity of resources, among these, the attention of others. This has naturally an impact on the status of knowledge.

5.3.2   The Acceleration of Everything: An Analytical Approach

The notion of space is now altered by the digital. The individual space, due to the ubiquity of digital artefacts. The contractual space, since contracts are traditionally defined by reference to geographical space (Simone 2012: 25).

Following Rosa and others, one can posit the hypothesis that the space-time of collective action is contracting, which poses a serious problem for organizations hitherto governed by verticality and control (bureaucratic organizations, or authoritarian governments).

Adopting a broad-brush anthropological perspective, we may be witnessing the emergence of a new kind of behavior (and a new kind of human being), linked to the ubiquity of digital and the accompanying acceleration.

In this light, digital objects and systems appear less like elements of infrastructure, and more like *boundary objects in the transformation of business and society, and of the associated modes of innovation and control*. The highlighting of the "experience" dimension in many of the projects—which, as the dictionary tells us, is all to do with "perception" and "sensation", and is the etymological cousin of "experiment"—is a very important dimension of the emerging mode of production. Digital is a continuous space of experience both for individuals (or customers) and for businesses and their strategies, which are conceived less and less in linear terms—designed to gain competitive advantage—and more and more in terms of engagement and experiment, and thus as a continuous process of trial and error, implementing the ideas that work.[2] There are similarities here with the shrinkage of space-time: the ever-shorter space-time of collective action, with unstable roles and organizational boundaries.

Acceleration emerges here as an essential transformational phenomenon, one whose impacts and implications are not yet fully understood. The acceleration of spatio-temporal links in particular calls for fundamental reflection about its nature, its future, and the associated risks.

---

[2](Cf. Mcgrath 2013).

5.4 The Articulation Between "Enterprise Production Space" and "Social Production Space"

The articulation of these two spaces is an essential datum, given the increasingly indeterminate nature of their borders. Historically, the industrial mode of production was instated by isolating productive time (and as early as the Middle Ages in Europe, according to historian Jacques Le Goff) which had become controllable by means of ad hoc instruments of measurement: clocks, timing machines, and the associated concept of "hours worked". Modernity is above all about controlling predefined spaces (the enterprise). It is now clearly established that in the current phase of "late modernity" the boundaries between these spaces are being eroded; a source of tension, but also of fundamental changes in the way things are "produced". This does not, of course, in any way announce the disappearance of factories, or the modes of production traditionally associated with them. But it does herald the ubiquity of productive time, correlated with the ubiquity of digital. This question, and in particular the issue of the equivalence of norms, can also be considered in the long view, from a sociological angle.

5.4.1 The Importance of the Equivalence of Norms

By "equivalence of norms" I mean the idea that, in simple terms, social standards should have an equivalent at the level of the enterprise. This is particularly true today, as is attested by the difficulties businesses encounter with Generation Y, which has developed digital uses and collaborative behaviors in social networks that it does not generally find in the enterprise. The equivalence of norms calls for managers—and particularly CIOs—to take an interest in emerging behavioral norms, in order to offer digital solutions and modes of governance appropriate to the new behaviors. More conceptually, the equivalence of norms calls us to think in terms of differentiated production spaces, whose tensions and extensions are described in the ISD projects (Fig. 5.1).

5.5 Postmodern Condition and Digitality

Lyotard (1979) provides an elegant analysis of the emergence of the postmodern condition, and the role of knowledge. The postmodern condition is contrasted with the modern condition, e.g. "the existence of a metadiscours referring explicitly using a particular grand narrative as […] the rational subject or worker, the development of wealth, we decided to call modern 'the science that refers to it to legitimize itself'." (Lyotard 1979: 7). From this angle, digital technologies can be seen as a source of "the great narratives" of societies. It could even be said that they

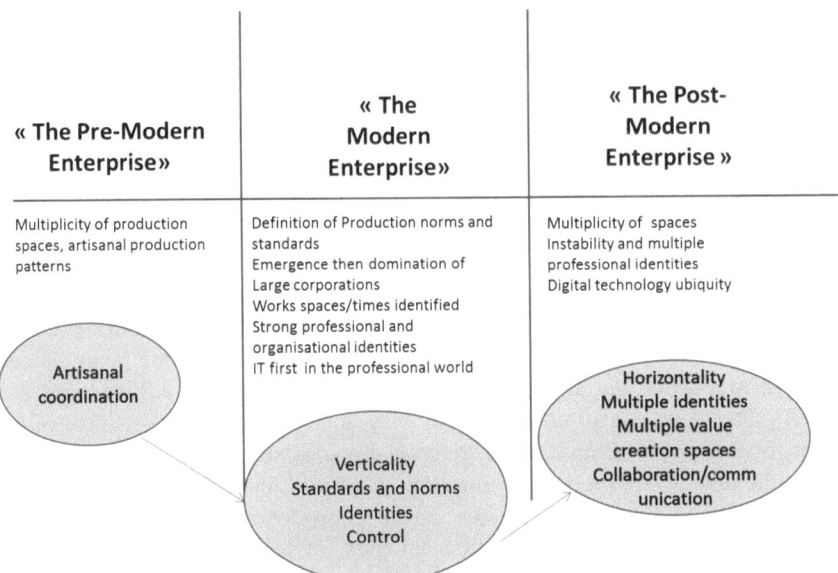

« The Pre-Modern Enterprise»	« The Modern Enterprise»	« The Post-Modern Enterprise »
Multiplicity of production spaces, artisanal production patterns	Definition of Production norms and standards Emergence then domination of Large corporations Works spaces/times identified Strong professional and organisational identities IT first in the professional world	Multiplicity of spaces Instability and multiple professional identities Digital technology ubiquity

Artisanal coordination

Verticality
Standards and norms
Identities
Control

Horizontality
Multiple identities
Multiple value creation spaces
Collaboration/comm unication

Fig. 5.1 Correspondence table of norms. A topography which poses the issues of equivalence of norms between the enterprise space stricto *sensu* and the spaces of the production of links

are "the coup de grâce unless we interpret 'digiworld' itself as the latest great narrative?" (Malecki and Moriset 2008: 222). Digitality can then be considered as both a source and the matrix for a hyper-postmodern society.

The postmodern condition and digitality can also be related to the issue of learning modalities and therefore intelligence. What approaches do postmodern societies take to learning modalities? Simone (2012: 32–37) argues that we are in the 'third phase' of the history of knowledge, i.e., how ideas and information are created. The first phase coincides with the invention of writing, which made it possible to present information on a stable support. The second began with the invention of printing, which allowed the large-scale production of knowledge via an ad hoc instrument: the book. These two phases are similar in the sense that they are both mental operations targeting texts. The third phase is mainly driven by videos and computers.

The invention of writing has had a huge impact of human beings, "it led to an immense increase in the importance of vision compared to hearing", and the emergence of a specific mode of perception: *alphabetic vision* (Simone 2012: 55). This has had an impact on the development of cognitive capabilities in modern humans. With the advent of digital technology we are moving from alphabetic vision (based on sequences), to non-alphabetic non-vision (based on simultaneity). Hence simultaneous learning and therefore 'simultaneous intelligence' has risen in importance (compared to sequential equivalents). This form of intelligence may be associated with a new type of human, the 'Homo Videns' described by Sartori

(1998, quoted in Simone 2012), in his analysis of the transformation of societies through the media artifact of television. People (especially children) lose their ability for abstraction, and therefore their ability to position themselves in space and time.

Digitality is therefore, at a minimum, a factor that is accelerating the emergence of a postmodern mankind—and therefore a postmodern enterprise, in which space and time are greatly compressed. As a result, relationships between people and aspects of the organizations are becoming more fluid. The question that arises is, if we follow the norms argument, to what extent does this instability call for a new forms of governance and institutions, and what new equilibria are needed between different organizational design principles and frameworks?

5.6 The Emergence of the Community Regime[3]

Developing a community perspective has become necessary because of the deep transformation of the implicit organizational order. Implicit order as used here reflects the fact that in many contexts people observe certain rules of behavior that are self-evident. From a long-term perspective, three types of order can be distinguished (Bounfour 2005):

- *Pre-industrial order*, e.g. the order that predominated before the industrial revolution. Generally, societies were governed by specific rules, and we can differentiate between a community (*Gemeinschaft*) and a society (*Gesellschaft*), as defined by the German sociologist Tönnies (1977). *Gemeinschaft* refers to an absolute unit where there is an indistinct and compact relationship between members. The perfect form of a community is the family. *Gesellschaft*, on the other hand, refers to a group of individuals who, while leaving together peacefully, are fundamentally separate. Under this regime, the individual is the center whereas under the community regime, the community represents the hub.
- *Industrial-manufacturing order* refers to organizational forms that were created and developed after the industrial revolution. Typically, while large hierarchies, whether private (e.g. General Motors at the beginning of the twentieth century) or public (e.g. the Department of Defense, Health Departments) emerged and developed, other organizational forms had already been tested and implemented including: communism in Eastern Europe, China and other parts of the world; clans in Japan; and market transactions worldwide. While each of these regimes has its own specific characteristics, overall, bureaucracies are more important than market forms.
- *Service-intangibles order*. This order corresponds to the present state of (knowledge) capitalism. There is one global socioeconomic system—the

---

[3]This section builds on my previous theoretical works on "communautalism". See Bounfour (2005, 2009).

'transaction regime'—that pressurizes every organization to continuously improve performance. Hence the increase in spot market transactions (the archetype being the financial market transaction) and the emergence of hollow corporations and networks.

Service-intangibles orders are linked to the potential of the Internet and the deployment of managerial practices brought about by the widespread deconstruction of social links. More importantly, and for the first time in modern capitalism, what drives innovation is not what is happening *intra muros* in vertical companies, but *extra muros* in societies and communities. This argument is developed further, and a distinction is made between different types of communities. The emergence of the community as an organizational mode of governance leads us to distinguish two regimes, the transaction and the community, in what has been called 'communautalism' (Bounfour 2006).

5.6.1   Two Regimes

From a systemic perspective, it can be argued that there are two parallel and potentially conflicting regimes: the transaction and the community.

5.6.2   The Transaction Regime

This remains the dominant model in modern economies. Companies and collective systems are driven by efficiency requirements, and therefore any individual or collective action is appraised from this perspective. In schematic terms, the return on invested resources is the alpha and omega of the assessment of any decision or behavior. Shareholder value is the archetype of such reasoning.

5.6.3   The Community Regime

The concept of the community regime was introduced following the deep crisis in societies with regard to 'recognition mechanisms'. It is clear that, at least since the mid-1970s, there has been a steady and profound trend towards 'fragilising' socioeconomic links within traditional vertical corporations. Outsourcing and networking activities, together with the emergence of the services economy have created a profound change in the way individuals see others and their organization and therefore, how they see themselves. This crisis—or transition—in recognition mechanisms is a useful perspective for understanding the overall dynamics of capitalism. From a microeconomic point of view, it can be argued that the market, and its transactions, is now the dominant form of activity. As a result, individuals have become orphans and seek new areas for recognition; hence the relevance of the concept of community. I defined a community as "*a set of individuals for whom relationships are governed to different degrees by 'recognition mechanisms'*". This naturally has nothing to do with 'communities of practice', that are claimed to exist within and around companies and organizations, as most of the time these so-called communities are in fact artificial groups that have been nominated or co-opted by management, without the necessary recognition mechanisms.

The concept of recognition was developed by Hegel (in his earliest work at Iena), and has recently been expanded upon by the French and German philosophers Ricoeur (2004), and Honneth (2002). It is central to the development of a critical theory of present and future societies, and therefore for building new social links within societies; in brief, for finding a new path to happiness. Finns are Finns because they recognize others as Finns, as do the French, Germans, Austrians, Japanese and others. They do not need 'communities of practice', as they form a natural community. A natural community is therefore a set of people who spontaneously recognize the others as fellows. Villages, cities, regions and nations are natural communities, shaped by history. In addition to these forms of communities, we can distinguish new forms that emerge from the transformation of large companies, and new forms of social-economizing that do not correspond to the traditional vertical form.

Three forms of emerging communities can be identified:

- *Constrained communities*, e.g. communities that individuals belong to because transaction costs are high. Typically such a regime can be—and is—applied to a network of experts or knowledge gurus. These individuals will exchange contacts, references, website citations and information in order to increase their market power. They do so because they are not able to, and do not have the resources or time to conclude contracts.
- *Quasi-Organic communities*. Although these communities operate under a transaction regime, they develop norms and behaviors that correspond to those defined by Tönnies. This particularly applies to the Linux community in the information technology arena, but also in other knowledge exchange communities. It can also apply to local communities such as districts, cities or villages.
- *Organic Communities*. These communities have yet to emerge. In organic communities individual expectations and behaviors, and the community's rules of governance are fully congruent.

5.6.4 Communities, Digitality and Intangibles

These theoretical considerations are necessary to understand the overall dynamics of the new production system, notably with respect to the importance and influence of digitality on the way these new forms of organizing are governed. Some scholars who have studied open source communities have even recommended a new form of organizing—the C-Form (Seidel and Stewart 2011). From this perspective, the community as an institutional order needs to be better documented, and its way of functioning must be analyzed (Marquis et al. 2011). This is, and will be a major component of emerging forms of what has been called the 'hybrid' organization (Haigh et al. 2015).

As we will show, this new form of organizing is necessary to understand how digital resources help to develop innovative offers, facilitate the flow of knowledge, and enable new experimental forms of entrepreneurship that are outside the traditional boundaries of large companies.

5.7   The Ethics of Use

The ubiquity of digitality and the immense potential of use of data resulting from it raise the issue of ethics related to such a use. This question of the ethics of use is closely related to the issue of dominant social norms, outlined above. In concrete settings, two very related issues can be put forward here : the first related to the general surveillance of people and organizations, and to what extent such a surveillance is accepted by societies and the issue of "commoditization of data" by firms and organizations, especially by major platforms.

With regards to the surveillance, the concept of liquid modernity has been extended to liquid surveillance: "how far does the notion of liquid modernity—and here liquid surveillance- help us to grasp what is happening in the world of monitoring, tracking, tracing, sorting, checking and systematic watching that is surveillance? The simple one-word response is "context"" (Bauman and Lyon 2013: p. 8).[4] We are in context where the issue of liquid surveillance is in the same time debated as well as accepted (as it is often mentioned the paradox of privacy). In concrete terms, firms, institutions, observe and watch most of the life of connected people all over the world. We witness practices and behaviors of superpanopticon type in many contexts and by major platforms (Poster 1995, in the case of data bases) and public institutions (the NSA case, see also the recent debate on the French law of citizens' surveillance). The question for managers is then: to what extent is it ethically acceptable to surveille behaviors of clients, users, personnel and others stakeholders? Are there specific rules to be observed? Is it even of interest of firms to adopt ethically acceptable and reliable behaviors? Is the system consistent enough or should we only assist to the generalization of "ethical washing" (as it is the case of green) behaviors in firms and organizations?

The second issue—the commoditization/monetization—of data is in part related to the first one. Some organizations collect data on everything regarding user's behaviors. These data are monetized especially in B2B relationships. Behaviors

---

[4]For a review of the studies on surveillance—especially the distinction between modern and postmodern approaches to surveillance, see Lyon (2007).

(such as search on google and similar platforms) are source of monetizable knowledge in publicity markets. These data can also be monetized in B2B relationship with suppliers. Here again, the question is the one of interest for firms and organizations to adopt a pure opportunistic approach related to the monetization of data: if data are undoubtedly a source of value, is there a need for the establishment of specific norms, regulations and standards for their valorization/monetization? Practices here are still emergent. But taking a more global perspective, we can foresee the adjustment of behaviors of firms to local norms of behaviors as well as to the more or less readiness of societies to the generalization of the panopticon practices and approaches.

In the research arena, we can observe recent initiatives, including in management science (The AIS Bright Internet initiative), trying to bring to the fore the importance of developing protection for privacy at the appropriate level, and calling for "a new and safer Internet platform, the *Bright Internet*, while protecting users' privacy at an appropriate level. The proposed principles are *origin responsibility, deliverer responsibility, rule-based digital search warrants*, and *traceable anonymity.* This endeavor requires the investigation of technologies, policies, and international agreements on which new business models can be created" (Lee 2015).

In the studies conducted under the ISD program, this question was addressed from the angle of work practices, but also by considering the point of view of public users in the Asian context. The research shows that the question of the ethics of use is central to the development of modes of production, especially when looked at in terms of data. For the professionals, the question is seen as important to the operation of businesses and organizations in general, but the mechanism used (codes of conduct, ethics committees, etc.) are usually deemed insufficient to handle issues of data use, including the proactive use of data by businesses. There are unmet expectations, therefore, when it comes to company practices with regard to digital uses, their foundations, and their evolution. In terms of public usage, the surveys carried out in Asia (China, Taiwan, South Korea and Japan), indicate that users seem to be more concerned by technical constraints on use than by the thought of operators tracking and using their data. This issue calls for further investigation. But one thing is certain: the question of the ethics of use has been very clearly raised by the ubiquity of digital, in its various dimensions. It is proposed as a determining factor in the composition of a specific scenario (*Resistance to digital*, in Chap. 11).

5.8 The Data Ecosystem

Data is not just a resource. It is an ecosystem that conveys and facilitates a new mode of production, in which the 'combinatory function' plays a central role. This is already the case in large digital platform-based companies (Google, Facebook, etc.), both in the West and the East (China). But the issue of data does not simply relate to visible platforms; it is also a question that faces every economic sector and every enterprise. For any governed organization, the indeterminate boundary between its activities and the mobilization of data represents both a source of value creation and a risk, even in apparently routine operational decisions (such as market research). Data is at the source of an ecosystem that is characterized by the abundance of links between information artifacts (the Internet of Things) and people, activities, and organizations. It therefore represents a potentially monetizable asset. When processed, with varying degrees of automation, it also provides new levers for the production of new knowledge in new ways.

5.9   A Synthesis: Five Key Dimensions

A thematic analysis of the propositions formulated by the program points to five key dimensions in the digital transformation of the enterprise: (1) the expansion and plurality of spaces; (2) the articulation between transactional links and organic links; (3) spatio-temporal structure; (4) organizational plasticity, and finally the central element of the analysis (5) the acceleration of links (Fig. 5.2).

5.9.1   The Expansion and Plurality of Value Creation Spaces… and the Transformation of Modes of Value Production

Value creation spaces have mushroomed with the advent of digital. The very borders of the enterprise have become fuzzy, and products and services are being developed in multiple spaces. As well as customers and suppliers, and "internal" resources—the traditional points of reference for business decisions—we must now

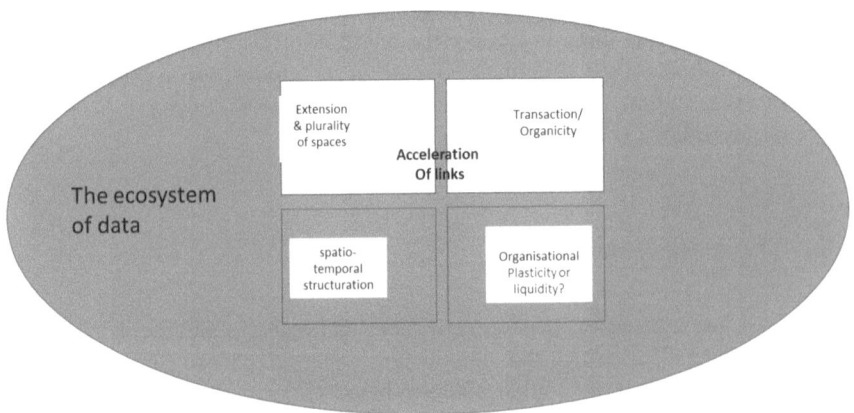

Fig. 5.2 The acceleration of links, at the core of digital transformation

reckon with complementors, mobility (of customers and of employees), the private time of company personnel, and the immense space of data.

Meanwhile, the nature of the modes of production of value is changing fundamentally. Traditionally, value is perceived (including in economic theory) as a process of consumption—and therefore literally destruction—of an output (of tangible goods in particular). In the digital economy and the digital society, value is created and deployed by multiple channels, including experiential channels (notably that of the customer experience). Value is created by an individual and collective experience. This experiential process further reinforces the non-directly-transactional character of modes of value creation. In digital spaces, the transaction is not directly linked to the underlying processes: the Google user, for example, is a resource rather than an object of transaction, but this attention resource is mobilized for a purpose which is indeed transactional, namely advertising. The Nike+ shoes analyzed by one of the projects, were designed as part of an experiential system, of which the end purpose is transactional, but the customer relationship here is more complex: the product/service transaction takes place in the context of a global production system in which the experience and the flow of information (behavior, benchmarking of individual users, belonging and contributing to a community of users) are essential to the actual realization of the transaction. The plurality of spaces is now compounded by the complexity of production systems.

5.9.2 The Articulation Between Transactional Links and Organic Links

From Ronald Coase onward, economists have recurrently theorized about the existence of the firm, as a resource articulated in a more or less organic way, and not by recourse to the market. Williamson developed the arguments essential to understanding the dynamics of resource organization, between market and hierarchy. The theory of organizations, building on the work of Chandler, justified the singularity of the firm and developed a historical perspective for analyzing its modes of configuration over time. Digitality, associated with a long deployment of new managerial practices (outsourcing, networking of activities, etc.) leads us to revisit the question of the relations between transactionality and organicity.

Transactions—market relations, in other words—are deployed massively in the internal domain of the enterprise, turning it into an organized market, but this deployment is limited by its cost and inefficiency, notably for empowering innovation. Innovation requires a minimum level of organic relations between individuals in order to spread fully throughout the company's internal space. Symmetrically, the market—and therefore the transaction process—becomes once again a key lever for organizing digital, including for one of the most complex of activities: innovation. It is now possible to get people of varying degrees of expertise to collaborate on an innovation problem without prior contact. In other words, thanks to digital, and the associated platforms, innovation can be organized around purely transactional links. If we extend the spectrum to all forms of social interaction, we find that organicity does not necessarily predominate; there is often a degree of hybridization between the two modes or "regimes". Generally, in the work sphere, interactions now combine transactional modes of governance with more or less organic modes of governance (dominated by recognition-based relations).

5.9.3   The Management of Space-Time

Control over space-time is a vital dimension in business management, one that is often seriously undermined by digital, which clashes with the traditional mechanisms in operation in the enterprise. Several projects describe how companies have revisited their way of doing things, especially in areas touching on product design, in geographically distributed contexts, and with broad functional implications. The KBO 2020 project, for example, illustrates the need for R&D-intensive enterprises to develop instruments for controlling spatial and temporal "alignment tensions" when developing products with strong time constraints. The Desvaldo project, meanwhile, shows how a global software development firm was led to strongly develop the skills of its Chinese partner, drawing heavily on agile methods, in order to cope with the constraints imposed by the geographical distance between its two co-development sites (one in the UK, the other in China).

5.9.4   Organizational Liquidity

By "liquidity" is meant the plastic (basically, malleable) character of the enterprise, its boundaries, and its activities. Liquidity is greatly facilitated by virtuality: "the virtual is by no means the opposite of the real. On the contrary, it is a fecund and powerful mode of being that expands the process of creation, opens up the future, injects a core of meaning beneath the platitude of immediate physical presence" (Levy 1998: 16).

This indetermination/plasticity is greatly facilitated and amplified by digitality, and by the generative nature of digital technology. The development of Cloud applications, crowdsourcing, and open innovation solutions, are just a few reflections of the huge move under way towards organizational liquidity. Its impact will have a determining effect on the 2020 enterprise.

5.9.5   The Acceleration of Links

The preceding elements prefigure an open production system, dominated by accelerated links between a plethora of value creation spaces inside and outside the enterprise. On the inside, the acceleration of links is illustrated by the wholesale deployment of mechanisms that shrink functional, disciplinary and geographical distances. On the outside, the massive use of market mechanisms to develop ideas and innovation programs also attests to the importance of the links between the different value creation spaces.

Chapter 6
From Lean to Acceluction: Complements or Substitutes?

The Lean production was the key concept proposed by the IMVP program (MIT) of the 1980s. The concept has been later refined into different variants, including the lean management. The Lean production refers to the way OEM optimize the tangible flux within the value chain of the automotive sector, especially between suppliers and OEMs. The research of the ISD program develops a new concept based on the articulation of links between different spaces of value creation. In the digital economy, the key concept is not the reference to the optimizing of transactions (as it was the case with the Lean), but the accelerated production of links, as the acceluction regime refers to it. The chapter delineates why and how we need to shift from the lean production towards a new production system, labelled here acceluction (e.g. accelerated production of links).

6.1 "Acceluction": The Mode of Production of Emerging Digital Uses

In light of earlier developments, should we now recognize the emergence of a new, digitally-driven, mode of production, and thus of business organization? If so, what are its key characteristics, and what are the implications for companies? The answer here is positive. The main arguments are developed below.

The characterization of the emergent mode of production is a key step in the design of the 2020 enterprise. Drawing on the work conducted to date by ISD, and more generally on the available literature on the transformation of socioeconomic systems and their modes of governance, we will attempt to characterize—in structural as well as conceptual terms—the emergent mode of production.

First, however, a brief review of some of the concepts relating to dominant modes of production, putting them in historical perspective. At the conceptual level, the modes of production that were, until recently, dominant were *designed* in one of the most widely studied industries—and certainly also the one that has generated the most management fashions and practices—namely, the automotive industry. At this level, the dominant concept over the last thirty years has been that of "*Lean production*", with its managerial variants (*Lean management*, etc.).

© Springer International Publishing Switzerland 2016
A. Bounfour, *Digital Futures, Digital Transformation*,
Progress in IS, DOI 10.1007/978-3-319-23279-9_6

6.1.1   Lean Production and the Space—Time Dimension

Let us pause for a moment on this concept and look specifically at its space-time dimension. The concept of "*Lean production*" grew out of a major MIT program in 1988 on the future of the automobile (the concept was mooted in 1988 by John Krafcik, a researcher at MIT (Krafcik 1988). Lean refers to a system put in place by Toyota in the late 1950s, after the organizational innovations introduced by Taiichi Ohno, and early 1960s: a centralized system for the streamlined production of a material good (in this case, cars) by optimizing flows from end to end of the process, from supply chain, to production, to customer relations (Womack et al. 1991: 48–63). Lean focuses on controlling material flows between industrial operators (suppliers and large manufacturers). The production space is relatively circumscribed, and is limited to the enterprises in the sector (including the distribution network), even if customer needs are integrated (Lean being a substitute for mass production). Lean production was developed by observing the dominant production system—the Toyota system—which all global carmakers, especially the American ones during the 1980s, sought to understand and to replicate.

Lean is a more advanced system than mass production, in as far as it optimizes material and information flows, including those that relate to demand, but in a relatively confined and controlled system (Fig. 6.1) consisting of the suppliers, the enterprise itself, and its customers (distributors).

The flow space here is pre-identified and in some ways relatively controlled: even if we are dealing with a so-called "extended" enterprise setup, the core

Fig. 6.1   The lean production space

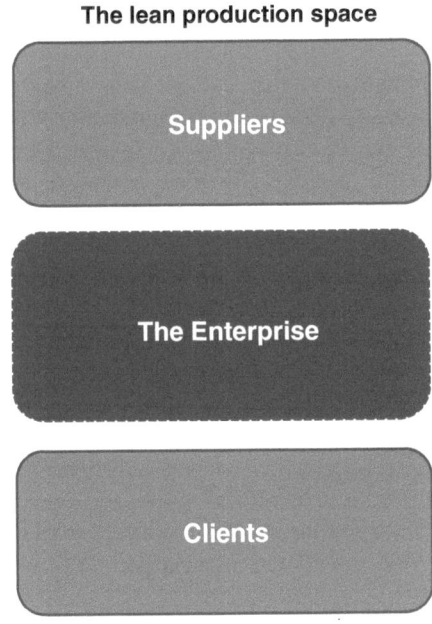

The lean production space

Suppliers

The Enterprise

Clients

Table 6.1 Lean production: key characteristics

Lean production (relative to mass production)
The principles of Lean production include: • Teamwork • A focus on communication • Efficient use of resources and elimination of waste • Continuous improvement
Compared to mass production, Lean production means: • 1/2 the human effort in the factory • 1/2 the manufacturing space • 1/2 the investment in tools • 1/2 the engineering time (hours) • 1/2 the new product development time • Less than 1/2 of the needed inventory on site

Source A table built based on Womack et al. (1991)

enterprises control and manage the production system, albeit with one important adjustment relative to mass production: the need to integrate downstream information flows (from customers). Despite the extension of the productive space, it remains relatively closed, but above all stable. Admittedly, information standards were introduced to the industry, in a pioneering way, by Toyota in the 1950s, but elsewhere only much later, in the 1980s, with EDI and the emergence of standardization bodies: Galia in France, and its European counterpart Odette.

Lean production is very much a production system governed by flow optimization (Table 6.1).

Let us consider closely how the concept of lean production is defined in the book of the IMVP by Womack et al. (1991). The book opposes the mass—producer (e.g. the Western producer) to the lean producer (e.g. the Toyota producer): "The mass—producer uses narrowly skilled professionals to design products made by unskilled or semiskilled workers tending expensive, single-purpose machines" (p. 13), whereas "the lean producer, by contrast, combines the advantages of craft and mass production, while avoiding the high cost of the former and the rigidity of the later. Towards this end, lean producers employ teams of multiskilled workers at all levels of the organization and use highly flexible, increasingly automated machines to produce volumes of products in enormous variety" (p. 13). So lean production differentiates from mass production by the profiles of the skills and the way it uses resources, but also more importantly by the way it search for their continuous optimization. As it is clearly and explicitly stressed by the authors: "Perhaps the most striking difference between mass production and lean production lies in their ultimate objectives. Mass producers set a limited goal for themselves—"good enough", which translates into an acceptable number of defects, a maximum acceptable level of inventories, a narrow range of standardizes products. To do better, they argue, would cost too much or exceed inherent human capabilities" (p. 13); "Lean producers, on the other hand, set their sights explicitly on perfection: continually declining cost, zero defects, zero inventories, and endless product variety" (p. 13/14). So using the ISD program language, Lean production is fundamentally a system of optimization of (mostly) physical flux within the production spaces in the automotive industry (suppliers, OEM and to a lesser extent customers). Such a production system has naturally on specific functions of the enterprise: factories, design, suppliers, and dealership. They have been the subject of specific developments in the book.

But with digitality, bolstered by the Internet, the production space is radically transformed, in its economic, managerial, and henceforth social dimensions. As we noted, the production space is now broad and unstable: there is a degree of indeterminacy in its boundaries and in the identity of its participants.

This phenomenon is amplified, at the sociological level, by the advent of postmodernity, one of the characteristics of which is a lesser willingness to heed rational, structured discourse and consequently to the forms of organizational order linked to the mass production phase. There is no causality between post-modernity and digitality (the former preceded the latter chronologically), but digitality has certainly amplified certain behavioral traits of late modernity, such as individualism, volatility, and skepticism (notably about the existence of stable structures). At the symbolic level, there is a displacement from value creation towards immaterial objects and signifiers, one of the digital variants of which relates to exposure in digital spaces.

6.1.1.1 End of Materialism, Beginning of Immateriality[1]

Another way to conceive of the structure of the world, and of human existence, is to consider the question from the viewpoint of the transition of technological systems along two axes: a "materials/energy" axis, which dominated the world in the 19th and 20th centuries, and a "man-biosphere relationship/structure of time" axis, dating to the beginning of the 21st century. The first axis is what Thierry Gaudin has called the "axis of materialism" and the second, the "axis of immaterialism". Following Simondon and others, Gaudin points to the decline of materialistic values such as domination, conquest and performance, and the emergence of immaterial values such as individuation, empathy and resilience.

In this context, the question—both theoretical and practical—that we must answer is twofold: (1) What are the arguments that justify the design of a new mode of production? and subsequently (2) What are its main constituent parts?

In what follows, I shall attempt to address these two points. I will restrict myself, for the most part, to ISD's field of expertise: the socioeconomic aspects of the enterprise.

---

[1]The development here is inspired by Thierry Gaudin's excellent presentation to the IC9 conference at the World Bank, for which I express my thanks: www.chairedelimmateriel.u-psud.fr.

6.2 Arguments in Favor of Recognising a New Kind of Mode of Production

Five arguments are put forward here once again to justify the proposal for a new mode of production:

- *The space-time dynamics of production spaces* (value creation), which go beyond the traditional scope of industrial analysis (customers, suppliers, partners), even if we use broader definitions such as "extended enterprise". Through the massive and ubiquitous development of digital uses, new spaces are appearing: the (hitherto) private time of employees, their mobile working time, the data space, the collaboration space and the social exposure space;
- *The multiplicity and instability of operator roles in the production space*: customers one day, suppliers the next, or perhaps partners in one-off or recurrent transactions;
- *The dynamics of link-acceleration at work explicitly and visibly in most spaces;*
- *The generativity of digital* as a technological system (more on this later);
- And finally, *the plasticity/liquidity of institutional and organizational boundaries*, which results from all of these developments.

6.3 Acceluction: The Central Concept that Characterizes the New Mode of Production

Schematically, taking a retrospective look at the question of modes of production, we can distinguish between three successive modes of production, each characterized by a struggle to take control of a particular resource:

- The agricultural mode of production, characterized by the importance of controlling the land;
- The industrial mode of production, in which one of the key resources is the work force, which needs to be controlled;
- The emergent mode of production, founded on digital, in which the key resource is fundamentally immaterial: it consists of increasingly accelerating links, which the economic operators seek to control. It is this mode of production that I have called **acceluction**, being a mode centered on the accelerated production of links. My aim in so doing was partly to signal the expansion of value creation spaces, and, more importantly, to recognize that value is now created by a rapid expansion of transactional or organic links (Bounfour 2005, 2006, forthcoming), and their subsequent acceleration.

Adopting a sociological perspective, and following Touraine (1973), a production system is determined by several elements of a technological nature, but also by various types of power relationship, and thus by the behavior of potentially organized actors (Fig. 6.2).

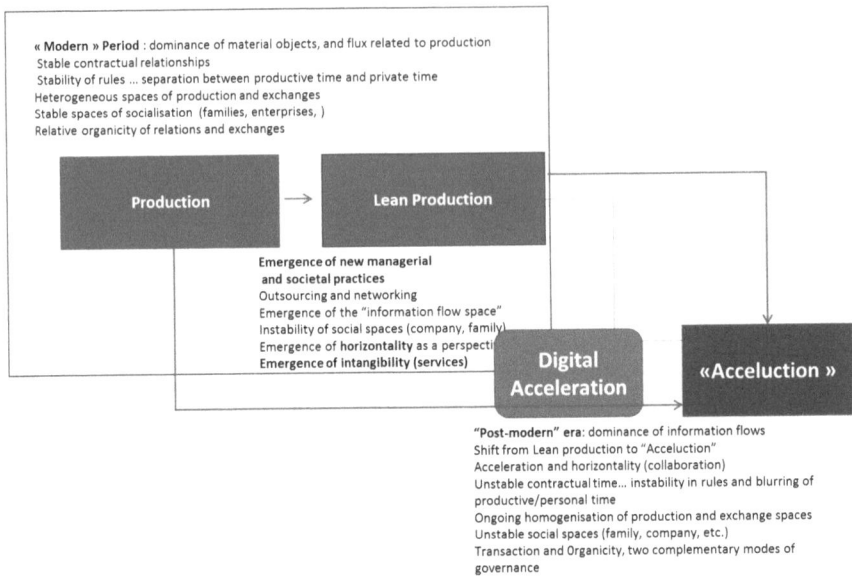

Fig. 6.2 Mass production, lean production and "Acceluction". *Source* Bounfour (2011)

6.3.1 Transactional Links

By transactional links, I mean links established either in a market, or in the context of an enterprise, but governed by market logic. Examples of the former include bids for patents, or the practice of open innovation (e.g. Innocentive in the USA, or Hypios in France) or the development of a market for exchanges within the company. Transactional links are not exclusive to collaborative forms of production.

6.3.2 Organic Links

Organic links point to spaces generally of the community type, or *Gemeinschaft* in the sense of Tönnies (1997), in which relations are governed to varying degrees by recognition. This principle usually governs communities of researchers, for example, or certain communities of open-source developers, or natural communities such as regions, towns, or villages. Organicity emerges as a significant perspective due to the crisis in *implicit order* (by which I mean the nature of the contract between individuals) in large organizations.

Digital links may or may not be organic: a link established militantly with others on a social network is, in principle, organic; a sale on eBay is not.

6.3.3   Topography of Acceluction

If acceluction is the emergent guiding principle for (digital) enterprise governance, then we must establish its topography. Simply put, acceluction aims to characterize the importance for the 2020 enterprise of mobilizing its digital resources, in order to articulate its links (both transactional and organic) with ***an extremely wide-ranging set of value production spaces***, as it encompasses markets (not just customers), organic (and less organic) communities, hybrid organizational forms, and society at large.

6.3.4   Acceluction and Digital Generativity

Generativity refers to a technology's capacity to produce sudden changes driven by a large number of diverse and uncoordinated participants (Zittrain 2006, cited by Yoo et al. 2014). It is contrasted with modularity, which tends to define problems in terms of predefined subsystems that can be controlled from a central point (such as a large company). The analysis of the spread of digital technology in recent studies by Yoo and his colleagues—sponsored by ISD—points to the generative nature of digital technology, and its transformational and in some ways unplannable character, as observed in APIs and mash-ups on digital platforms. This demonstration of the generative nature of digital dovetails, at another level, with the importance of links between actors and technological building blocks in specified digital spaces. The spatio-temporal analysis of generativity demonstrates the value of developing a

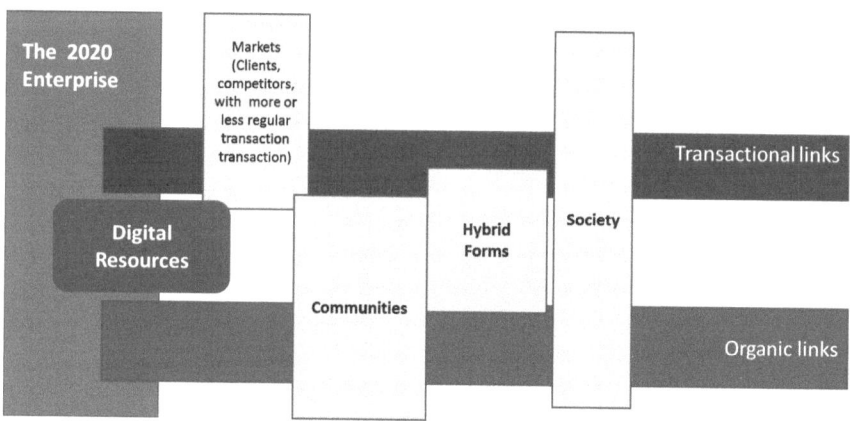

... A large scope of action for the Enterprise in mobilising its digital resources

Fig. 6.3   Topography of acceluction. *Source* Bounfour (2011)

dynamic, outward-looking vision of digital innovation: what Yoo and his colleagues (Yoo 2014) call: "the generative digital platform" (Fig. 6.3).

This topography sets out an immense scope of action for the enterprise (and its CIO), in mobilizing its digital resources. Hence the importance of the equivalence of norms between the enterprise's space of governance *stricto* sensu and the spaces where links are generated.

Chapter 7
The Liquid Enterprise and Digitality

Bauman proposed the concept of liquid modernity/the liquid society to characterize current societies. Similarly, enterprises, through the support of digital resources, are developing various degrees and forms of liquidity. The chapter develops the concept of the liquid enterprise through an examination of its different dimensions—time, space, coordination, planning of resources, contracts, etc., and how this concept is consistent with that of the liquid society. Here, the focus is on the importance of the consonance of norms and behaviors. As most of the digital transformation is beyond the sphere of control of firms, it is important for them and their managers to build their approaches and tools upon societal rules and norms. The chapter delineates the content and modalities for such congruence.

7.1 Congruence and the Preeminence of Societal Changes

Bauman developed a very interesting perspective on how societies function and how their members are related and interact. The 'liquid' concept refers to the way society is no longer considered as a structure but increasingly as a "matrix of connections and reconnections which are the results of the hazard and a number of possible permutations, by essence infinite" (Bauman 2007: 9). In his book, Bauman develops arguments about how societies are organized and governed, and the implications of new norms for behavior and individual strategies.

The arguments we examine here relate to the collapse of reflection and collective action, and as a corollary the fragmentation of lives, which do not include ideas of "progress" or "career". Hence, the game is refocused on individuals rather than structures; consequently individuals are required to become more flexible, rather than respect predefined rules. From this perspective, society is no longer an organic structure, but rather a collection of individuals who are searching for a high level of flexibility in interactions.

© Springer International Publishing Switzerland 2016
A. Bounfour, *Digital Futures, Digital Transformation*,
Progress in IS, DOI 10.1007/978-3-319-23279-9_7

7.2   From Liquid Society to Liquid Enterprise

According to the principle of the equivalence of norms, liquidity in social inter-actions is also observed at the firm level, although it is difficult to identify its origins. At the societal level, postmodern behaviors are often dated to the mid-1970s. At the firm level, restructuring, especially in the West, became wide-spread in the 1980s. The restructuring of firms that has taken place over the past three decades, and the associated impacts on employees' "mental space" has con-tributed to their increasing distance from the strategic and organizational discourse of management. The emergence of outsourcing (which started in IT, when Kodak awarded a contract to several services providers in 1987) together with the emer-gence of the networked enterprise (where Cisco has identified as an emblematic example) are other sources of the liquefaction of enterprises. Liquidity in this case, is not only—or even primarily—linked to digitality; instead it is closely related to managerial practices, which have progressively led to the liquidity of firms and therefore to the liquidity of social contracts.

7.2.1   Generation Y as an Illustration

An illustration of liquidity is the results of a series of workshops organized during preparations for the ISD program, with the so-called generation Y. Twenty par-ticipants were asked about various topics: their fundamental personal values, their relationship to their employer's strategy, working relationships, how they saw their career developing, their view of digital objects, how they spent their time (both at work and elsewhere), and finally how they saw the future of their employer.

It emerged from discussions both in focus groups and plenary sessions that their relationship to their employer appears to be almost transactional, i.e. governed by purely rational calculations. In most cases, these young people behave according to a 'win-win' principle, illustrated by the way they spend their time: if responding to an email sent by their boss in the evening takes 3 min, they will spend the same 3 min on social media during working hours. This is an illustration of the liquidity of firms that is consistent with the general principles defined by Bauman for society. Furthermore, when it comes to career development, this generation applies the principle of the "impairment test", similar to International Financial Reporting Standards for specific intangibles. They continuously test the value of their skills in the job market, even if they are not looking for a job. Naturally, digital artifacts facilitate this behavior. The results of the workshops suggest that while such behavior is indeed facilitated by digital artifacts and systems, it also finds its origin (at least for some participants) in the waves of restructuring that their parents had to face. These events created the roots for liquidity that now translates into specific interaction behaviors within and around firms.

7.3   The Liquid Enterprise and Digitality

In their book on the economic history of industrial revolutions, Freeman and Louçã (2001) describe the components of the most recent Kondratiev wave: the ICT revolution. They describe its main ingredients as computers, telecommunications, institutional (regulatory) settings and a new organizational design based on the networked enterprise. If we consider these ingredients to be the major components of the new production system, liquidity and digitality are key aspects, and are closely related to the ongoing transformation of our economies and societies. We can then state that the liquid enterprise and digitality are closely related phenomenon, without necessarily establishing a causal relationship between them.

Here, the liquid enterprise refers to plasticity in both its modes of governance and its boundaries and resources. Digitality is great enabler of such plasticity: it contracts the time-space of the firm, and facilitates the accumulation of resources without boundaries or even significant investment. Consequently, it is a serious challenge to the social contract found in modern organizations, particularly the so-called 'salarial' (traditional) contract.

7.4   Liquid Enterprise, Liquid Management

A liquid enterprise calls for liquid management. This raises the issue of aligning managerial practices with liquidity requirements. In terms of business strategy, this means that it should be thought of as a succession of decisions that are adjusted to market and local conditions. As for investment, decision making is a continuous trade-off between external and internal resources that are permanently under pressure. With respect to collaboration and coordination, incentive systems must take greater account of the intrinsic nature and behavior of people, especially the fundamental nature of their social contracting process that is increasingly governed by liquid relationships.

7.5   The Liquid Enterprise and Organizational Design

The liquid enterprise and liquidity in general are interesting concepts related to the future design of enterprises and societies. The implications are explored in detail in the chapter on acceluction. Here, we simply underline that liquidity is a key component of the future design of firms and organizations. It can be argued that theoretically, as far as the design of organizations is concerned, there are limits to liquidity and a form of solidity is needed for collective action. However, the nature of artifacts may lead to specific behaviors where speed and acceleration are the major drivers for performance. In this case, especially in organizational contexts that are dominated by intangibility, liquidity may represent a major advantage for socio-economic organizing.

Chapter 8
Acceluction: Stakes, Opportunities and Risks

The acceluction mode is challenging for companies, their executives and their IT managers. The chapter presents the high stakes, opportunities and risks, related to it, in term of business development, resources reallocation and contracting. The chapter analyses into details the different dimensions of acceluction, its fundamental tensions and the related stakes, opportunities and risks. More generally, the chapter develops the arguments towards moving from a management centered on developing "one best ways" to a management centered on "tensions", underlined by the acclucted mode.

8.1 Acceluction and Digital Strategy

At the design level, acceluction requires companies to consider three key issues:

1. *Identifying value creation spaces*: an essential starting point for understanding the dynamics of the digital revolution, and for understanding acceluction as a production system. In addition to its own space, the enterprise will need to integrate multiple spaces governed by a range of different principles: customers, competitors, complementors, social networks, and society in general;
2. *Determining the types of links to develop* with the enterprise and with the actors in these spaces: transactional and/or organic links, but also more or less specialized links, with new implications in terms of contracts and legal liability, especially as regards intellectual property rights;
3. *Defining a strategy and an overall governance structure for the topography of links*—acceluction also calls for the definition of a general governance structure for links, in other words, for extremely fine-grained management of the wealth and multiplicity of links in various defined spaces. This is conducive to the development of an architectural approach, facilitated by the company's digital resources.

The acceluction model calls for a fresh approach to the question of digital and value creation in the enterprise, in its three dimensions. Table 8.1 below lists the strategic and operational questions that correspond to these three dimensions.

© Springer International Publishing Switzerland 2016
A. Bounfour, *Digital Futures, Digital Transformation,*
Progress in IS, DOI 10.1007/978-3-319-23279-9_8

Table 8.1 The digital strategy founded on acceluction: key points

Identifying value creation spaces	List the spaces and actors concerned. Evaluate the nature and stability of their relations. Determine how they articulate with the enterprise and outside the enterprise. Develop an overall approach to the entire ecosystem for the 12 spaces identified
Types of link to develop	From the analysis of the ecosystem as a whole, determine the types of links to be developed with each space and each associated actor: simple market transaction (commercial) links; information and knowledge exchange links (transactions on social networks); links representing structured relationships based on a certain level of trust (co-developing products in sectors with a program logic, or exchanging reputational data with structured communities attached to the enterprise); short term links versus long term links; proprietary links (e.g. on exclusive data) versus joint links (shared data)
Governance	Identify the internal and external stakeholders. Determine who does what in terms of managing identified and validated links. Evaluate the overall coherence. Decide on a dynamic control mode for digital links and updates (e.g. for data monitoring). Define the management instruments (strategic value, economic value, associated risk)

8.2 The 2020 Enterprise: Its Underlying Tensions

In sum, the enterprise of 2020 will be an accelucted enterprise, whose business models and overall governance will be centered on the management and accelerated production of multiple links, constantly renewed. Digital resources are already, without any doubt, essential levers in this process.

The analytical elements presented here serve to define what I propose to call the *acceluction regime*. The acceluction regime is a regime of tension between two differentiated regimes: the *liquidity regime*[1] and the *solidity (or organicity) regime*.

Figure 8.1 sets out the key criteria.

These tensions were revealed explicitly or implicitly in all of the ISD projects. Now for a closer look at each dimension:

8.2.1 Liquidity-Plasticity/Solidity-Organicity

The liquidity-plasticity of the 2020 enterprise refers to the volatile and unstructured character of its value creation and resource spaces. Liquidity-plasticity can be seen as one of the key dimensions that make up the 2020 enterprise. It is reasonable to suppose that liquidity spaces will mainly be governed by transactional links, while

---

[1]Borrowing the terminology of Bauman in *Liquid Modernity* in the first Wave A report (Bounfour 2011, CIGREF Foundation).

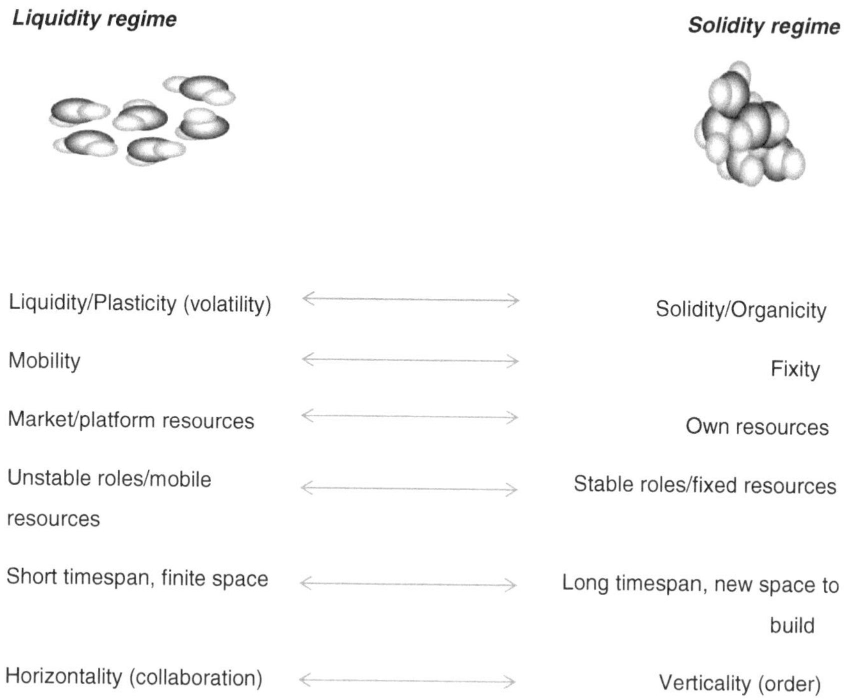

Liquidity regime   **Solidity regime**

Liquidity/Plasticity (volatility)	⟵⟶	Solidity/Organicity
Mobility	⟵⟶	Fixity
Market/platform resources	⟵⟶	Own resources
Unstable roles/mobile resources	⟵⟶	Stable roles/fixed resources
Short timespan, finite space	⟵⟶	Long timespan, new space to build
Horizontality (collaboration)	⟵⟶	Verticality (order)

Fig. 8.1 The accelution regime and Organizing 2020: between liquidity and solidity. *Source* Bounfour (2013). Nb: images for liquidity and solidity are from Wikipedia: https://fr.wikipedia. org/wiki/Liquide

solidity spaces will, in large measure, be governed by more organic links. In any event, the 2020 enterprise will have to articulate its liquidity spaces with its solidity (organicity) spaces.

8.2.2   Mobility/Fixity

As the ISD program has shown, mobility is an essential space in the design of the 2020 enterprise. The 2020 enterprise will have to manage tensions between its mobile resources (enabled by mobile information artifacts such as smartphones or other future devices, including those of its employees) and its fixed resources— installations, factories, offices: the spaces of design, production and distribution. Mobility is one of the major spaces of the 21st century. Its generalization in public and private use is bringing about a fundamental transformation in ways of designing, producing and distributing. Fixity, *a contrario*, remains an important dimension of organizing, including in its interaction, socialization and innovation components. Enterprises will put particular emphasis on the first dimension, with mobility becoming a generalized front office; this is already the case for Samsung, which aims to get 70 % of its personnel mobile within 5 years. Others will clearly follow, as mobile information artifacts become a key lever of acceleration.

8.2.3 Market Resources-Platform Resources/Own Resources

The digital revolution inevitably raises questions about the deployment of the company's technological, human and information resources. The 2020 enterprise will need to interface three types of resource: market resources, mobilized mainly by means of transactions; platform resources, developed jointly with others; and is own, proprietary resources, controlled by the enterprise. The recent development of innovation platforms, which can safely be expected to intensify over the next ten years, offers prospects for transactions on previously unimaginable knowledge markets. When it comes to allocating resources, the enterprise will have to make trade-offs between three types of decision:

1. *Acquiring resources on transactional markets* (innovation resources such as crowdsourcing, information resources such as cloud data);
2. *Developing resources—especially digital resources—on shared platforms* (sector platforms, service platforms);
3. *Developing its own resources.*

These three types of decision are obviously neither new nor original items on the business manager's agenda. But their implications are greatly amplified by the ongoing digital revolution. The potential costs and benefits are now of a different order: choosing market resources poses problems as to their articulation with other resources (platform resources, internal resources); selecting collaborative platform resources can generate new opportunities, but also new vulnerabilities; and the types of internal resources to be developed have to be considered with a view to their dynamic complementarity with other resources. This raises issues of agility, of strategic positioning (choice of platforms) and, in many contexts, of risk and vulnerability (including intellectual property rights), as well as of what stance to adopt in external community spaces, depending on the specific enterprise context.

8.2.4 Unstable Roles, Mobile Resources/Stable Roles, Fixed Resources

The instability of roles is one of the major impacts of the digital transformation. By role, I mean the enterprise's position in the productive system: customer, client, supplier, competitor, etc. Traditionally, this position is defined within the industry's value chain. But with the digital, the concept of value chain—and, for that matter, the concept of industry—is challenged by the instability of roles. In the era of mass production, and even of Lean production, these roles were generally seen as clear, identifiable, and relatively stable. Acceluction integrates role instability: customers can be suppliers and competitors at the same time. *Positions are transitory*. As a result, there is a crucial problem with the mobility of resources and thus of investments. For the 2020 enterprise, integrating the instability of roles, and thus the mobility of resources, is an essential condition for success.

8.2.5 Short Time-Span, Finite Space/Long Timespan, New Space to Build

The accelerative dimension of digital has been underlined as a key component of the new paradigm about to be constructed. The contraction of space-time forces us to consider the movement of enterprises in a short timeframe: managerial action is ultimately reduced to a succession of short timespans, and thus to a finite space-time that is constantly being explored. For the 2020 enterprise, this is certainly an essential dimension, one that highlights the constantly adaptive nature of actions in digital spaces. While this perspective is now dominant, is does not exclude the consideration of a long time span, that of the space to be constructed, and consequently the future sequence of short timespans to be explored.

8.2.6 Horizontality-Collaboration/Verticality-Order-Hierarchy

With Web 2.0, horizontality emerges as a central trend in the organizing of the enterprise. By horizontality is meant, in simple terms, a way in which people work collectively and thus a way in which action is organized. Horizontality can be associated with collaborative forms observed in communities, where work gets done through engagement, ability, persuasion and seduction. Relational capital also plays a major role. Verticality, by contrast, plays on notions of hierarchy; the legitimacy of action—and consequently of decision-making—is generally based on hierarchical empowerment.

In today's digital world, and even more so in the world of 2020, this latter approach is challenged; by generational practices, but also as a result of the importance of social networks, whose governance is closer to horizontal forms of cooperation than to vertical forms.

The ISD studies show that this trend is not found evenly and unambiguously across all enterprises; in the broader picture, it is all about context, organizational culture, the core business... in other words, contingency. The work on Generation Y also invites a more nuanced analysis, in particular as regards relations with the enterprise and the way projects are managed. For the 2020 enterprise, the future outlook is one of tension between verticality and horizontality, rather than a uniform horizontality.

In any event, the 2020 enterprise will need to find the best possible articulation between these two dimensions of organized action.

The governance of the 2020 enterprise—an "acclucted" enterprise—leads us to examine the tension between two regimes: the solidity regime and the liquidity regime (Table 8.2).

Table 8.2 The digital strategy of the acclucted enterprise and its underlying tensions

Underlying tensions	Concrete management situations
• Mobility/fixity	• What business models for mobile uses in the enterprise? • How can we turn mobility into an effective front office? • How do we redeploy staff from fixed-status tasks to mobile statuses? • What are the appropriate coordination mechanisms?
• Market resources/platform resources/own resources	• How do we make trade-offs between own investments (legacy, internal teams) and market resources (cloud, etc.)? • How do we manage the question of intellectual property rights and digital assets (proprietary rights, shared rights)? • How do we deploy digital platforms, and how do we manage the data issue? • How can we integrate external spaces (social networks)? • How do we handle the question of joint resources (including employees)?
• Unstable roles, mobile resources/stable roles, fixed resources	• In partner relationships (customer, suppliers, etc.) how do we assess the stability of roles and statuses (will customers always be customers)? • What stability is there for investments, and so for the return on investment for resources allocated to our customers and partners?

(continued)

Table 8.2 (continued)

Underlying tensions	Concrete management situations
• Short timespans, finite space/long timespans, new space to build	• What is the timespan of the company's activities (R&D, design, innovation)? If the timespan is short, which particular tensions need to be identified? What coordination mechanisms should we deploy? • What is the (short-term) impact of acceleration on the company's working climate (stress, employment relations, etc.)?
• Horizontality-collaboration/verticality-order-hierarchy	• What incentive systems for Generations C and Y? • How to tradeoff between horizontality and verticality? What role can the central functions (general and functional departments) play? • What is the right incentive system? How do we define its control (time, task, and performance monitoring) and what adjustments are required? • How much tension is acceptable to our employees?

The *solidity regime* is the paradigm around which most managerial discourse—particularly with regard to large companies—has been constructed over the last century, and especially since the emergence of the modern enterprise: organicity of interactions, fixity of resources (including personnel), community regime, work on long timespans/(vertical) spaces to build, and specialization of resources. But, as we have seen in several contexts (and this is not the least of digital's impacts), the solidity regime—and with it that of the *solid enterprise*—is fundamentally subverted by the liquidity regime.

The *liquidity regime*—and consequently that of the *liquid enterprise* is characterized *a contrario* by a set of rules: transactional regime; mobile resources (including personnel) and unstable roles; short timespans and finite space; market or platform resources (see developments in Chap. 7).

The acceluction regime is a regime of tension between these two (sub-) regimes. The experiments conducted under the ISD program clearly illustrate this tension: the solutions trialed or rolled out by companies are sometimes liquid in character, sometimes organic. The emergent mode of production, that of acceluction, does not indicate that digital enterprises will necessarily be liquid enterprises; what it indicates is the importance of this liquidity.

8.3   The End of the "One Best Way"... and the Regime of Permanent Tension

From what we have seen, it is clear that the management of the 2020 enterprise will be based on permanent trade-offs between options with contradictory terms. The ubiquity of digital makes the behavior of actors, and therefore of enterprises, highly unstable and thus very open. As several of the ISD projects have shown, with concrete examples, enterprises are henceforth governed by a regime of tension (between partnerships, functions, geographies, timescales, etc.). With the multiplication of value creation spaces and of potential positions, there is no longer any "one best way" that all enterprises should aim for. There are only separate, unique paths. This evaluation criterion puts decision-makers in a far more uncomfortable position than before, due to the open nature of the game, the uncertainty, and the risk attendant on any position they choose.

Chapter 9
The Acceluction Regime: Its Governance

The acceluction mode calls for an adapted governance for the 2020 digital enterprise. The chapter develops the general principles of governance, especially with regards to the different stakeholders. It specifies the nature of tasks to be achieved, the breakdown of responsibilities as well as the coordination mechanisms. The chapter also details the new role of the digital function within and around enterprises and organizations.

In this chapter, I will consider the value spaces of the acclucted enterprise, before detailing the issue of governance structures and approaches.

9.1 The 2020 Enterprise: Its Value Creation Spaces

In previous chapters, I have defined the conceptual building blocks for the emergent mode of production, around the concept of acceluction. How, in concrete terms, is it deployed?

Based on the analyses conducted by the program, we can examine five types of space that need to be articulated:

- *The traditional space of Lean production*: suppliers, internal resources, customers;
- *The social space*, a space now contiguous with the company's productive spaces, including—importantly—the data space. Within this space we find three value creation sub-spaces, employees' private time, social networks and entrepreneurial space;
- The space of *Competitors, Complementors and Platforms (C, C & P) resources*;
- The *mobility* space;
- The *data* space.

Figure 9.1 represents the topography of these spaces and their mode of articulation.

What is striking here is the extraordinary expansion in the space of the enterprise, especially for large companies, and the vital importance of paying attention to the interstices between the spaces and the company's overall digital strategy.

© Springer International Publishing Switzerland 2016
A. Bounfour, *Digital Futures, Digital Transformation*,
Progress in IS, DOI 10.1007/978-3-319-23279-9_9

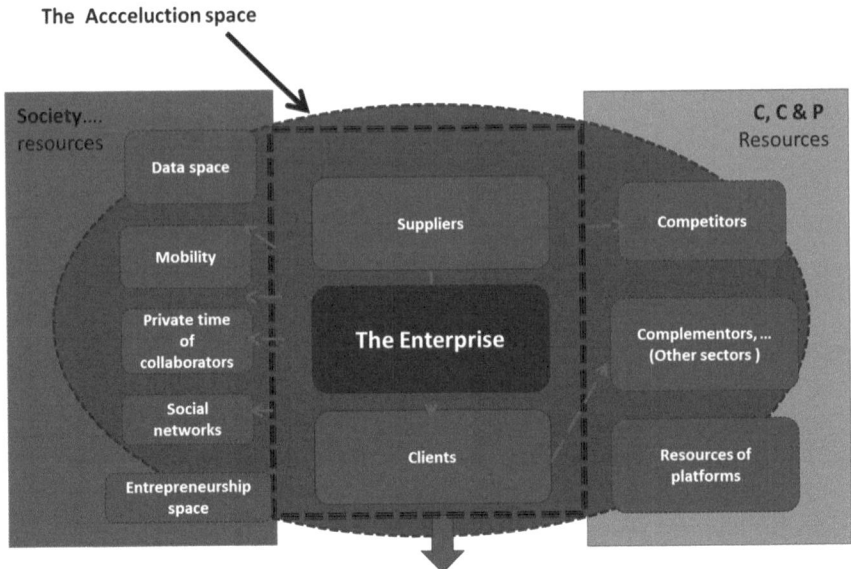

Fig. 9.1 Lean spaces versus acceluction spaces. *Source* Bounfour (2013)

9.2 Value Spaces and the Governance Issues

From a global perspective, governance has to do with institutions. As De Soto
(2001) elegantly demonstrated, institutions are important components in economic
growth. This was underlined by the World Bank (2006) in their study of the wealth
of nations. Institutions also played an important role in the industrial revolution
(Verley 1997). However, in the context of post-modern firms that have been highly
influenced by the digital transformation, there is a need to question their role,
especially in the context of organizational liquidity. What are the key ingredients
for social exchanges (Baudrillard 1972, 1981) in a context of hyperdigitality? How
do we govern digital spaces in a context where the principle of self-organizing may
be more appropriate than top-down approaches? To what extent is the generativity
of digital technology associated with new governance structures? These profound
questions are beyond the scope of this book. At this stage, we simply consider
general approaches to governance, before analyzing aspects of governance in the
acclucted enterprise.

9.2.1 General Approaches to Governance

The paper by Drnevich and Croson (2013) provides an integrated theoretical view
of the role of IT and business strategy. It synthesizes five theoretical perspectives

and their key concepts, profit mechanisms and integration with management information system (MIS) research: collusion/coordination, governance, competence and flexibility theories. The following discussion of the issue of the governance of digitality builds on this work.

Here we consider governance-based theories and IT technologies and discuss their relevance for our understanding of the dynamics of digital spaces.

9.2.2 Governance-Based Theories and Information Technology

Governance approaches are based on transaction cost economics such as those developed by Coase (1937) and Williamson (1975), and agency theories (Alchian and Demsetz 1972; Jensen and Meckling 1976). Two economic profit models are associated with these core theories: the Ricardian theory of profit (operational efficiency), and Coasian transactional rent efficiency (avoiding transaction costs).

These theories provide the foundations for the trade-off between internal resources and those which have to be sourced from the market. Such ideas have been mobilized in various IT contexts, notably outsourcing and off-shoring decisions (Bounfour 1999; Hirschheim and Lacity 2000).

These theories provide a stimulating perspective for defining the boundaries of firms' activities. However, in practice their implementation has proved difficult, due to the problem of calculating transaction costs and acceptance of the overall approach in decision making.

Transaction and agency costs are only one side of the story, and we need to integrate/revisit the fundamental nature of the firm, through an examination of its skills and capabilities.

9.2.3 Governance-Based Theories of the Acceluucted Enterprise

Theories of governance are based on a dual approach: the firm and the market. From a geographical perspective these are two fundamental spaces. They are also based on transactions, e.g. discrete activities that can be measured in monetary terms. This is how economic theory traditionally measures value: it requires a transaction in the market that can be measured in monetary terms.

However, as we have shown, value is now created in multiple spaces—the firm has become boundaryless and is surrounded by multiple, instable spaces. Nevertheless, it is governed by market mechanisms—some hybrid, others fully organic. Therefore, it is difficult to delineate the boundaries of the 'firm', based only on principles related to the efficient allocation of resources.

Furthermore, the dimension of time, and the 'acceleration of everything' must be carefully considered. Rent instability and volatility in digital spaces raises questions about the relevance of calculating transaction costs in many cases.

Finally, transactions that are expressed in monetary terms only tell one side of the story, and links are often more important. This is illustrated by this expression often heard in digital communities, "Ubiquity first, Revenues Later" (URL).

9.3   Governance Structures for the Accelucted Enterprise

Based on the foregoing, what are the principles that firms and organizations (especially large ones) should consider when developing and implementing a governance structure that is in line with acceluction requirements?

Several components must be taken into account: general principles of governance, governance bodies, suitable leadership, fostering an ad hoc culture, and reconsideration of processes.

9.3.1   General Principles of Governance

The general principles of governance can be derived from the key characteristics of acceluction as a new production system emerging in digital spaces. Four fundamental principles can be identified:

1. The importance of links as a major source of value creation in digital spaces;
2. A re-assessment of the scope of value creation, especially its platformic/ecosystemic dimension;
3. The integration of the time-space dimension as a critical component of digital strategies;
4. The importance of aligning the firm's culture and processes with these three fundamental principles of organization and management.

9.3.2   The Governance Agenda

These four principles must be translated into ad hoc action programs in the context of the governance agenda of large firms and organizations. The agenda should be directly supervised by the company's general management who coordinate action with both internal and external stakeholders.

Next, we examine the five components of the digital governance agenda for the accelucted enterprise (Fig. 9.2).

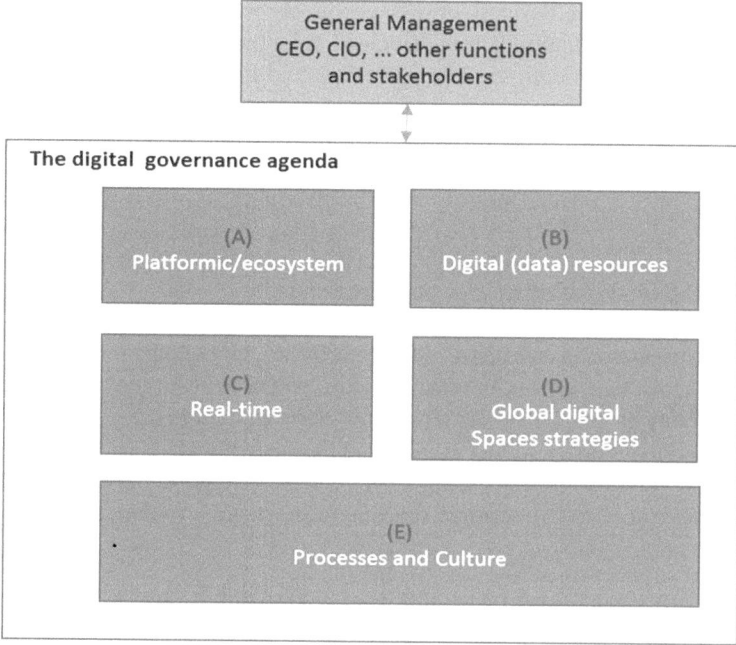

Fig. 9.2 The digital governance agenda for the acceluced enterprise

9.3.2.1   A-Strategies and Programs for Platforms and Ecosystems

The platformic form has become the dominant way to organize digital spaces. Platforms are critical for the continuous existence of any business, including those that are most 'protected' (the public sector). Every sector and every enterprise is affected and challenged by the platformic form. One speaker at a recent conference on digital societies indicated that in the near future Google will be able to raise taxes. Therefore, there is a need for every organization to develop strategies and programs that specifically target the plaformic form of value creation. Various positions are possible: a firm can develop a central position within a given platform, or can contribute as a service or data provider to another. In either case, there is a need to rethink platforms as key spaces for digital strategies, whether globally or in very specific geographic zones (for instance clusters, or knowledge-intensive cities and regions). Platformic strategies and programs must therefore be considered as an issue that is closely related to the development of innovative ecosystems.

9.3.2.2   B-Strategies and Programs for Data/Digital Resources

Data are and will be a major source of growth for companies and organizations. Their use and the modalities of their leverage will impact the way firms develop and even survive. The core business of every company is now challenged by the need for data leveraging, sharing and valuation in different contexts. Therefore, in parallel with platformic programs, companies have to develop ad hoc strategies and programs for data management, especially data that is considered as a strategic digital asset. To achieve this, it is critical that firms develop a transversal view and establish new programs for data processing, in order to ensure consistency in terms of positioning, value and actions.

9.3.2.3 C-Real-Time Strategies and Programs

Time and space are the two sides of the coin for the digital transformation of firms and ecosystems. Real time is an issue for management in post-modern societies, due to the volatility of needs and behaviors. It is also important because of reduced time spans related to products, services, links, and business models. Real time is therefore also an issue for the digital agenda of the accelucted enterprise, but does not exclude the development of a long-term perspective for navigation in digital spaces.

9.3.2.4 D-Strategies and Programs for Global Digital Spaces

Digitality underlies a profound change in the geography of firms (especially large corporations). Digital artifacts are more and more deployed globally, which necessitates revisiting coordination mechanisms in units that are spread all over the world. One such mechanism might be to empower local units and increase their knowledge capabilities in such a way that they can make a greater contribution to the overall innovative offer of the firm.

9.3.2.5 E-Strategies and Programs for Processes and Culture

Finally, for all of this to work it is necessary to align processes and culture with objectives for each of the identified components (platformic, data, real time, and space strategies). In practice, this means that processes should be lean, limited in number, and transversal—firms need a lean COBIT for their digital strategy to succeed. There is also a need to change the firm's culture, notably by developing an overview of the issues and understanding why they are critical. This must be accompanied by openness to these new approaches and any related and necessary risks.

9.3.3 Leadership

The accelucted enterprise calls for effective leaders who are highly adaptive (Bennis 2013). This is necessary in order to response to the time-space requirement. Leaders must be ready to learn from failure in specific contexts in order to strengthen the firm's resilience, which entails "coming back effectively and rebounding from difficulty and adversity" (Bennis 2013: 636). Leaders need to understand—and be able to allocate the necessary time—to grasping the impact of the digital revolution on their business. They also need to look closely at the managerial implications in terms of speed, scope and risk. In short, they need to develop an overall roadmap for their organizations, which incorporates both daily adaptive behaviors and the long-term exploration of new ventures.

9.3.4   Governance Bodies

It is not particularly useful to discuss the details of who should do what in organizations, as there are multiple functional and organizational contingencies. Furthermore, this role is the responsibility of managers. Although the ubiquity of digitality makes it the business of every member of the organization, it is expected that general management (the CEO and CIO, and others such as chief data officers and chief platform officers) will play a key role. The most important factor is the alignment of their functions and activities with the five major components of the digital agenda.

Chapter 10
From Data to Digital Assets

Beyond a general discussion of big data as the new frontier for business growth, it is important for firms to develop an analytical approach, notably with respect to digital assets. Based on previous arguments, the chapter proposes a categorization of digital assets, their value and risks, and their mode of approach, taking into account the overall governance structure.

10.1 Background: The Added-Value of IT Artifacts and Systems

Creating value from IT has been widely discussed in the literature and various approaches have been developed (e.g. consumer surplus, productivity, complementary and intangible assets) (Bounfour and Epinette 2006). The paper by Nicolas Carr published in the Harvard Business Review (Carr 2003) challenged IT value creation, calling it a "commodity". The paper provoked a lively debate on the value of IT. The author recommended executives to invest less in IT and develop a "follower" approach to such investments. This controversial argument is particularly contentious if we take into account the intertwining of IT and the firm's assets. In many case, investment in IT is transformational, as it relates to critical strategies and policies, such as innovation, product development, global coordination, etc. Therefore the idea that IT is a commodity is debatable when it comes to actual projects and settings. In some cases, internal IT departments are even more efficient than the market, *ceteris paribusi*, not to mention transaction costs that are related to contractual arrangements with the market and follow-up.

Given the ubiquity of digitality within and around firms and organizations, the scope and scale of the debate has changed dramatically. It no longer concerns the level of investment in specific, identified and localized IT functions, instead it is about investing in a new source of growth and defining its control modalities: data. The question of whether IT matters or not has become irrelevant; it has been replaced by a discussion of the modalities of leveraging an asset that represents one of the firm's major investments. This issue is the main focus of data-driven strategies and innovation.

© Springer International Publishing Switzerland 2016
A. Bounfour, *Digital Futures, Digital Transformation*,
Progress in IS, DOI 10.1007/978-3-319-23279-9_10

10.2   Data-Driven Innovation as a Perspective

Here we examine the issue of data-driven innovation, building on a recent OECD interim report (OECD 2014). This report discusses data-driven innovation and its contribution to economic growth. For the OECD, this contribution is made along two channels:

- Via the specific properties of data as an *infrastructure resource*, which can be leveraged as input in the production process, and therefore enhance productivity;
- Via *value creation mechanisms* based on data analytics, which can be used to gain insight and therefore increase knowledge, or automate decision processes.

These two levers are complementary ways to assess the contribution of data to innovation and growth on a national scale. The report also provides recent estimates of the size of the market and firms, especially in the rapidly-growing ICT sector.

The report highlights that the market for data analytics is already substantial, and anticipates that globally it will grow from USD3 billion in 2010 to USD17 billion in 2015. It also considers the significant challenges posed by data leveraging for economies, in terms of both demand and supply. From the demand side, the major issues relate to the development of skills and organizational change within the firm, a point that was highlighted in our discussion of organizational design (Galbraith 2014, quoted in Chap. 1). Supply side issues relate to investment in broadband, data access, and cloud computing. Finally, from the societal point of view, the major issues relate to market concentration, loss of autonomy and freedom, security, inequality and change in job markets.

The stakes and opportunities for data-driven societies appear to be very high. An important question relates to how firms can take advantage of these changes. There is a need to develop frameworks and tools that can assess the potential for data-driven innovation and value creation.

10.3   Data and Value Creation

In many firms and organizations, data are considered to be an essential lever for digital transformation and value creation, now and in the next decade. Although several authors have emphasized the importance of big data for firms (Davenport 2014; Mayer-Schönberger and Cukier 2013) the issue is not really new. It has been the subject of many applications—notably in marketing—particularly in the context of decision modelling and statistical analysis. Recent research has underlined the importance and superiority of data-driven decisions for performance (Brynjolfsson et al. 2011). However, beyond the massive quantity of data that is available, the real novelty lies in the multiplication of combinations of data, especially with a view to value creation.

The importance of data also relates to emerging production systems, particularly platforms, which have been greatly facilitated by the generative nature of the digital revolution, as shown by some of the ISD program's work. Moreover, in an accelerating world, the real-time use of data becomes a key variable for decision making. Finally, social networks have generated huge quantities of unstructured data that companies, in general, make no use of.

10.3.1   Why Think in Terms of Digital Assets?

One of the central challenges for managers lies in the evaluation of data as a digital asset, and therefore in the assessment of its economic value generally. There is a need to shift the narrow focus on data to reasoning in terms of digital assets. We should underline here that an asset is a resource controlled by a company via specific property rights. Data is a digital resource that can be seen as a joint asset (i.e. something that is, or can be shared with others); hence the importance of analyzing its status in terms of control, potential value creation, and risk.

10.3.2   The Issue of Connecting Revenue to Data

In the digital economy, linking data to revenue generation is a major issue. As an intangible resource, data must be bundled to ensure revenue generation. This process necessitates the definition of an overall strategy for data pricing and value generation: for example, can it be sold to external partners in a specific format? Or should it be leveraged by the firm itself in targeted marketing strategies? Or can both options be adopted? In any case, policies and programs need to be clarified at this level, with clear guidelines and principles regarding privacy, and more generally ethical use.

The Data Valuation Modelling (DVM)® approach to data valuation.

The Data Valuation Modelling approach is being implemented in several contexts.

The aim of the model is to bridge the gap between the expectations of executives—CIOs in particular—of data value creation, and what is actually available in the market. DVM is consistent with the most advanced state of the art in leveraging data for value creation, in a context where it is abundant.

10.3.3   Next Steps?

What are the next steps in the valuation of data as digital assets? According to the *Data Valuation Modelling*® (DVM) approach, five specific steps are recommended:

1. **Listing and categorizing the status of data as digital assets**
   This step is consistent with the general principles derived from overall trends that have been identified, the structure of the future enterprise and its digital transformation. Here the objective is to classify data into specific digital assets, notably those that are proprietary to the firm, and those to be created jointly with other members of the ecosystem.

2. **Analyzing the level of maturity of firms *vis à vis* specific steps and practices**
   For each asset (or generically) a set of practices are defined, which are assigned a level of maturity. Ethical issues and the alignment of general policies and practices are considered here.

3. **Assessing value in monetary terms**
   For each asset (e.g., data held on a loyalty card in a general distribution channel), we define the components of value creation and allocate or simulate revenue. Based on this step, a value can be attributed to the asset.

4. **Defining the firm's overall data strategy in terms of combinations**
   Firms allocate investment to assets. Revenue is generated from assets and their bundling. In this step, firms and organizations discuss the actual and potential level of value generation based on present or future/selected combinations of key digital assets.

5. **Developing simulations and scenarios for data valuation**
   Based on all of these elements, managers and their teams can develop data value simulations. This exercise may lead to the identification of scenarios where there are high returns, while others might be weak candidates for monetization. This step is essential in closing the gap between managers' expectations (and speculation) and what can really be achieved by data-driven value.

Chapter 11
The 2020 Enterprise: Six Contrasting Scenarios

Finding what companies will turn into by the year 2020 is rarely the object of such academic work as that delivered by the ISD programme. The traditional concept of Lean is based on constant improvement of the model through the exploitation of the company's potential. By 2020, the book suggests it will be replaced by acceluction, which is a wider analysis based on the multiplicity of defined value spaces. This multiplicity also means new tensions such as long term versus short term or company's own resources versus market resources. The identification of these tensions conveyed by the acceluction regime enables the author to formulate six contrasting scenarios for the design of the 2020 enterprise.

The actual deployment of the design components of the 2020 enterprise, as set out above, needs to be modulated by taking into account the context of the evolution of digital at the global level, where the factors of influence are social, technological, legal and organizational. The aim here is not to embark on a large-scale prospective study, but simply to propose a set of pointers to give us a clearer picture of where digital is heading, and the transformations it will bring about in the enterprise.

The 2020 horizon chosen by the program is both near and far. Five years, after all, is almost twice the average timeframe for corporate strategy plans. This exercise can therefore usefully be combined with companies' strategic reflections on their digital transformation.

This chapter defines the selection criteria for the scenarios. It then goes on to list them and to analyze their key characteristics.

11.1 Definition Criteria

The research conducted under the aegis of the ISD program has revealed a series of trends, but also a number of tensions which can be expected to have a major influence on the way digital is deployed in businesses, and thus on their level of transformation. Taking up the five dimensions defined by the program, the determinants of this deployment are strategic, social, organizational, technological and regulatory. Two of these areas—strategic and organizational decisions—lie within

© Springer International Publishing Switzerland 2016
A. Bounfour, *Digital Futures, Digital Transformation*,
Progress in IS, DOI 10.1007/978-3-319-23279-9_11

the discretionary power of the company, but social, technological and regulatory changes are outside the company's remit, even if companies can, and often do, influence such changes. It is therefore reasonable to assert that the speed of digital deployment in the enterprise will be determined by the extent to which the company accepts its implications. Particularly so when the issue of personal data usage arises, notably in a market environment. Clearly, also, the sometimes rapid deployment of information artifacts (the Internet of Things), with their capacity for dialog and "collective intelligence", will affect companies' digital choices. Finally, as a corollary, the way regulatory texts are drafted and transposed into national or regional statutes, will have an impact on the level of deployment of managerial practices, especially those relating to data, or to the critical issue of copyright.

Each scenario focuses on a particular dimension, indicating the implications for the digital governance of the enterprise.

11.2   Scenarios

A set of representative scenarios is presented below, before going on to analyze their implications for the profile of the 2020 enterprise.

11.2.1   Six Characteristic Scenarios

The identification of the tensions conveyed by the acceluction regime enables us to formulate six contrasting scenarios for the design of the 2020 enterprise (Fig. 11.1):

Scenario 1: *Polyspaces*
Scenario 2: *Back to basics (a return to the integrated enterprise)*
Scenario 3: *Mesospaces*
Scenario 4: *Platforms rule*
Scenario 5: *Network abundance*
Scenario 6: *Resistance to digital*

Fig. 11.1 Six scenarios for the future of the digital enterprise

1 Polyspaces	2 Back to basics	3 Mesospaces
4 Platforms rule	5 Network abundance	6 Resistance to digital

Scenario 1: *Polyspaces*
Dominant: Markets and organization of the enterprise

This scenario extends existing emergent phenomena and amplifies them. It implies that the enterprise is just one actor among many, and that as value creation spaces are multiple, the enterprise, in this environment, will need to coordinate (predominantly external) resources. This scenario speaks to the fragmentation of value creation spaces and their instability. In this context, digital ensures coordination between the multiple value creation spaces (data, mobility, customers, suppliers, complementors, etc.).

Scenario 2: *Back to basics (a return to the integrated enterprise)*
Dominant: Markets and organization of the enterprise

This scenario is, in certain aspects, the opposite of the first. It posits that the multiplicity of value creation spaces is problematic, due to the large transaction (and therefore governance) costs that it entails. Above all, it sees organicity as a key success factor for any entrepreneurial activity. In concrete terms, it assumes that the enterprise will redefine governance rules, and contracting rules, with a long-term orientation, and not founded solely on transactional calculations. Here, the enterprise regains the lost legitimacy, both among its employees (renewed trust) and with its external stakeholders. In this context, digital is a key lever for internal coordination and for transactions with stakeholders.

Scenario 3: *Mesospaces*
Dominant: Organization of the enterprise, territories and public action

This scenario supposes that value is henceforth created at the intermediate levels (collaborative platforms, networks, communities, regions, territories), with the local dimension predominating. It also considers that it is consequently at this level that the organizational models introduced by digital should be developed. This scenario outlines a certain territorialization of value creation, which may potentially have a strong geographical dimension, but may also concern organizational forms that are not territorially embedded. Specific social networks, for example, do not necessarily have a clear territorial attachment.

Scenario 4: *Platforms rule*
Dominant: Market structures

The main argument here relates to the structure—and in particular the infrastructure—of digital markets, at the global level. This scenario posits the reinforcement of the current oligopolistic structure of the digital market around a small number of players. Globally, we are witnessing a bipolarization of digital infrastructures, into a Western pole, dominated by the USA, with a handful of players (Google, Facebook, major Cloud operators and the like), and an Asian pole dominated by China. Platform-based structures are developing around "natural monopolies" whose market power is being reinforced partly by the generative character of digital technology—this specificity, in terms of the shaping and diffusion of innovations, has been demonstrated by the ISD program—and partly by the market power of the incumbent operators, who have sufficient financial resources to enable them to internalize any threat of external innovation from any entrepreneur. Under this scenario, neither the competition rules (notably in Europe), nor social norms and values (use of personal data) can prevent the big players flexing their market muscle.

Scenario 5: *Network abundance*
Dominant: Technological, social, regulatory

This scenario is defined using the criteria of network abundance as specified in the M2Mod project: the three Versus (volume, velocity, variety) of data; connectivity at the edges; autonomic networking (self-awareness) of digital objects; and extreme customization. Network abundance goes beyond the framework of the Internet of Things, as it includes networks of people. This scenario therefore focuses on the exploitation of Big Data by operators, associated with the Internet of Things. This dual movement makes it necessary for operators to manage significant tensions: (1) the tension between the autonomy of objects ("fly-by-wire") and decision support; (2) the tension between security and privacy; (3) the tension between data ownership and profitability; (4) the tension between public goods (data as a public good that everyone can use) and private goods (the appropriation of data by economic actors, including businesses).

This scenario assumes that most of the identified tensions are resolved, notably with the autonomization of objects and a shift from exchange-centered value design towards experience-centered design. Firms such as Wal-Mart or Amazon are expected to benefit fully from this scenario, including by offering services hitherto provided by other players (such as insurance), thus successfully leveraging the concept of customer learning.

Scenario 6: *Resistance to digital*
Dominant: Social and regulatory

The key dimensions in this scenario are social and regulatory. Here, the question of data usage becomes central, and resistance arises to a form of digitality. The battle for personal data becomes a social issue, but the regulatory texts remain fragmented and heterogeneous. Social networks become a massive space of expression for the protection of personal data.

This scenario assumes that certain tensions are not resolved, particularly those around data usage. Legislations remain heterogeneous in the major zones of the world economy (North America, Europe, Asia), as well as at the World Trade Organization. As a result, businesses cannot exploit the full potential of data ownership.

This scenario rules out a shift to an experiential mode of value. The traditional models continue to exist: car sales, traditional insurance, low decision autonomy of digital objects.

11.2.2   Scenarios and Profile of the 2020 Enterprise

For each of the scenarios thus defined, we can characterize the most appropriate profile for the 2020 enterprise (Table 11.1).

Table 11.1 Typical scenarios and profiles of the 2020 enterprise

Typical scenario	Profile of the 2020 enterprise
1—Polyspace	Acceluction governs the development of links between multiple value creation spaces. The 2020 enterprise will have to select the types of links it needs to develop, and subsequently manage them. Links can be transactional (giving rise to monetary transfers) but can also represent the exchange of information and knowledge. Open innovation practices, crowdsourcing, and the use of markets reach maturity. At the social level, the current heterogeneity of norms and rules enables enterprises to capitalize on links with these spaces (including user data)
2—Back to basics	In this scenario, the 2020 enterprise refocuses its activities internally. The practices of outsourcing, massive use of the Cloud, and more generally of the market, reach their limits, for reasons due mainly to the importance of human capital and trust as a factor in the growth and prosperity of the enterprise. In this context, acceluction is expressed in two ways: internally, via the acceleration of business and management processes (R&D, innovation, IT…), and externally, via coordination with partners, the securing of market positions, and the development of the customer experience
3—Mesospaces	In this scenario, the 2020 enterprise focuses its acceluction efforts on developing links between spaces of different statuses (networks, communities, territories, enterprises), often with a local or territorial base, but sometimes with a global reach (such as large science campuses like Saclay). These links are both transactional and organic. The variety of statuses and structures is a key dimension here
4—Platforms rule	This scenario stresses the platform dimension of digital. Under this scenario, the 2020 enterprise finds its digital strategy heavily dependent on its own platform strategy, as well as on its links with the major digital platforms, which dominate and capture the lion's share of the value created. Innovations are organized and articulated around these global-scale platforms. The platform structure becomes the hub around which most digital activities are organized. Acceluction plays an important role here in links between platforms and their client/partners, as well as with users (establishing profiles, service offerings, etc.)
5—Network abundance	Under this scenario, the 2020 enterprise harnesses the full potentialities of digital, notably through strong connection between physical and virtual spaces. This scenario assumes that all legal, social (privacy) and technical uncertainties have been resolved regarding the availability, circulation and monetization of data. It is in this scenario that acceluction acquires its full meaning. Value is created by the acceleration of links between physical objects and virtual spaces, between different business data (proprietary, joint, non-proprietary), and between (and within) businesses. The digital spaces thus defined become essential levers of digital transformation. By leveraging network abundance, enterprises amplify the growth of their business

(continued)

Table 11.1 (continued)

Typical scenario	Profile of the 2020 enterprise
6—Resistance to digital	This scenario assumes a kind of social boycott of digital uses, or at least a certain form of data-centric digital use, leading large sections of society to reject it. By resistance we mean here heightened vigilance about the use of personal or work-related data in contexts where there are few controls. On top of this there is, at the international level, serious fragmentation of the legal framework, creating uncertainty that prevents uniform use of data globally. Under this scenario, the 2020 enterprise develops a strategy attentive to the ethical dimensions of use and their effectiveness. It co-constructs solutions with users, and shares the rent with them transparently. It also develops vigilance mechanisms on intellectual property aspects, in a fragmented legal context

Chapter 12
Beyond 2020: Network Abundance, Data, and the Future of Organizing

What does the future look like for enterprises post-2020? To address this question, and building on the outputs of the ISD program, this chapter refines the factors that lead to change and disruption in the accelucted enterprise. This includes the emergence of networked abundance that goes beyond the Internet of Things. More generally, this chapter redefines the future as an asset for enterprises; it examines the impact of the various dimensions of digitality and how they will impact the design of the future of enterprises.

Post-2020, as the M2Mod project underlined (El Sawy et al. 2014) it is likely that the emergence of massive data will have a significant influence on the design of enterprises. The authors raise a number of questions, notably:

- Given the volume of data that is collected, will enterprises have to update their business models on an ongoing basis?
- How can an enterprise be managed in real time, in the context of abundant networks?
- Will it be necessary to dedicate human resources to data processing, or can algorithms be developed to process the data that is collected automatically?
- How can we train employees in the integration of new data ecosystems?

More generally, in the context of the perspective developed here, these questions raise the issue of how to develop an organizational design approach that is suited to a purely digital enterprise?

We briefly consider these managerial issues, before discussing their societal counterparts.

12.1 Managerial Issues Related to Post-2020 Digitality

Let us return to three of the questions raised by El Sawy and his colleagues (2014).

© Springer International Publishing Switzerland 2016
A. Bounfour, *Digital Futures, Digital Transformation*,
Progress in IS, DOI 10.1007/978-3-319-23279-9_12

12.1.1   The Question of Decision Making

Fundamentally, organizations and firms are decision-making 'machines'. In a situation of network abundance (machine-to-machine interactions in particular) the question of decision making at the strategic (the design and choice of business model) and operational level arises. Who will decide what? How to define and implement decision making? How to manage risks related to decision making? How to develop quality assurance principles and what should be their content?

12.1.2   The Real Time

Here again, the issue is about organizing activities, decision making and coordination. In this context, data analytics will play a critical role; organizational processes will be aligned with real-time requirements.

12.1.3   The Need for Specialized Human Skills

Data abundance requires the development of new skills, especially in data analytics. The availability of high-powered algorithms will create demand for such skills. The emergence of new profiles will challenge existing resources, particularly if they are ill-suited to real time and network abundance requirements.

12.1.4   The Future Organizational Design

Underlying these questions, we can glimpse an outline of the *post-2020 digital enterprise*—an enterprise in which the central issue is data: how it is processed and monetized. Post-2020, this change will clearly require an ad hoc research agenda focused on organizational design.

12.2   Societal Issues Related to Post-2020 Digitality

The foresight studies that were reviewed in Chap. 2 raise several issues related to the future of our societies and global systems, notably with regard to the impact of digital technologies. Based on these issues, and the six scenarios we have identified, several issues emerge which merit consideration from a global perspective.

12.2.1 Forms of Social Interaction

In the anticipated context of widespread machine-to-machine interactions, what is the role of human interactions? How do the three spaces (Ba) distinguished by Nonaka and Konno (1998) articulate: the physical Ba, the mental Ba and the virtual Ba? Will the virtual Ba become the standard space for social interaction? Will there be bottlenecks? What will be the impacts on the three major activities of mankind: work, leisure and learning?

With respect to work, several questions arise: What will be the new forms of work? How to create trust among people? Will the knowledge worker form become generalized? What is the future for Taylorist activities such as call centers? What about artisans and craftsmen and the "shop class" (described Crawford 2009)? Will there be hybridization of tasks? What kind of renewal in the nature of tasks?

With respect to learning, we have already witnessed the fact that traditional learning processes are under pressure, including at university level. Will we have gone beyond MOOCs to develop new forms of learning by 2030? Will campuses and universities remain the dominant forms of learning for younger generations? What about companies? What about individuals? Increased life expectancy will require the continuous renewal of human capital, especially in a context where freelancing and individual ventures may become a major feature of socioeconomics, and therefore of revenue generation.

With respect to leisure (and travel), business models are already being fundamentally revised, with new players entering the market on a daily basis. In the arts, for instance, we can expect the widespread development of digital resources in museums. More generally, digital resources will amplify the current trend of the sharing (accessible) economy, especially in a context where the world is moving towards more collaborative forms of regulation (the fusion scenario presented in the CIA foresight study).

12.2.2 Intangibility and Digitality

The question here relates to the type of exchange instruments used by people, especially in a context where acceluction becomes a major production system. Due to the multiplicity of spaces for value creation and the ubiquity of digitality, we can expect exchange and social interaction to become organized along intangibles such as brands, data, and reputation. We can also expect traditional forms of knowledge to become digitized and therefore more easily disseminated worldwide (an example is the way the Massai café, as community product, has been branded). At the global level, we can expect to see the emergence of collective goods such as collective brands, or collective knowledge that is relevant to specific communities and is widely disseminated via digital artifacts. In this context, joint IPRs will become a major lever for knowledge dissemination. Finally, as can already be observed in many spaces, monetizable and non-monetizable assets will coexist in different value spaces (markets, networks, communities, territories).

12.2.3   The Status of Employment and Job Opportunities

The OECD Development Center Foresight Study stressed the issue of job creation at the global level, due in particular to the impact of technology. The topic is rising to the top of the agenda for policymakers and citizens, especially given the limited number of jobs created by platformic digital firms, compared to the level of employment in traditional multinational companies. Google, Facebook, Twitter are large companies in terms of market capitalization; however, they are not large companies in terms of employment. Most digital startups begin with a limited number of employees. More generally, the impact of robotization and the automatization of processes (including machine-to-machine interaction) raises questions about job creation by 2030 in absolute terms, particularly for the younger generation.

Forms of employment are also relevant in this context. What types of contracts will be offered in 2030? Will freelancing be the dominant mode of work? Will most workers become entrepreneurs?

Intergenerational links must be examined in this context. Will there be enough organicity in a context of hyperliquidity? Is there enough space to make these two terms compatible?

12.2.4   The Future of Institutions

Modern institutions—states, regions, cities, and firms—face the challenge of digital ubiquity. Traditional firms (and ultimately all firms) must deal with newcomers in the digital world. States, especially in the West—are suffering from debt burdens and lack discretionary financial power. They are challenged by large platforms, notably in how they exercise their main source of power: collecting taxes. Regions, cities and territories may cope better, due to their ability to cluster knowledge resources (for at least for some them) and therefore collect taxes. Nevertheless, in the medium term, digital platforms remain a challenge: they possess resources and talented human capital, and they can internalize innovations encountered in their environment. Digitality means that they can even collect taxes or carry out cross-border police surveillance. Does this mean that the Hobbesian sphere of influence will move from states to platforms? This remains an open question that merits further attention. There is a need for further thinking about the future of institutions and how they are designed in order to best-suit the next generation of organizational forms.

12.2.5 The Platformic Issue: China Versus the United States

Data are the new source of soft power in the new capitalist society. Illustrated by the acronym 'Gafa' (Google, Apple, Facebook, Amazon) these digital platforms are a key source of global power for the United States. If China protects its internal digital market and develops its own infrastructure (Alibaba, Baidu, etc.), it is naïve to think that this is only (or primarily) to exercise control. Instead, it is fundamentally due to the importance of platforms in the new global context. Europe (following the decline and takeover of Nokia), Japan, and the Brics (India and Brazil in particular) have not succeeded in building their own platform infrastructure. Therefore, the future will depend in particular on the way Chinese platforms are internationalized and the global scenarios that could develop. As Chinese platforms become international, their power will increase. In Europe, digital platforms might emerge if, for instance, privacy becomes a critical factor for differentiation in the eyes of citizens. Nevertheless, it is important to avoid developing a linear view of the future; digital spaces might yet hold great surprises, as innovations emerge from unexpected areas and unexpected regions.

12.2.6 The Status of Large Enterprises

As Chandler underlined, the large enterprise emerged from the need to internalize and diversify activities. In the context of the generalization of digitality, what does the future hold for them? Will they remain the final step in the growth of a company? Or will we see more ecosystemic forms of organizing, where small firms play a critical role? How will "micromultinationals" (described by Hal Varian), ecosystems and digital platforms interact? All of these questions remain unanswered.

Epilogue

The Future as a (Digitized) Asset

Unless there is a massive rejection of digital artifacts, our future will be digital. Digitality is a production system that is fundamentally transforming our daily life, relationships and every organization's position in space and time. As a result, any thinking about the future must be closely related to the use of digitality. The nature of this 'technology' creates many uncertainties, especially in terms of societal and economic relationships. Hence the importance of developing an agenda that is specifically dedicated to the future as an asset.

It is necessary to reflect on the future as an asset because of the expected fundamental change in many aspects of decision making, both for individuals and societies in general.

The acceleration of everything has had an impact on decision making and resource allocation in all organizations. Two examples are the business plan and terminal value in finance: these two instruments must be revisited according to the way we represent the future, and its uncertainties resulting from the impact of continuous change.

For individuals: their future is related to their capacity to build their own (mostly intangible) assets (expertise, skills, reputation, and personal brand). The transformation of ecosystems and the higher status given to freelancing requires individuals to rethink their future according to the hypotheses they create from their socio-economizing spaces.

For societies: the question of the future is an acute. In addition to the global issue of resources (climate, energy, water, food), there is the question of how to redesign social contracts in a context of hyperliquidity. Here again, the future is uncertain, and it merits a very systemic and rigorous examination.

© Springer International Publishing Switzerland 2016
A. Bounfour, *Digital Futures, Digital Transformation*,
Progress in IS, DOI 10.1007/978-3-319-23279-9

Annexe A
The CIGREF Foundation Governance and Activities

CIGREF set up the **CIGREF Foundation** in **2008** (under the aegis of the Sophia Antipolis Foundation) with a mission—*to better understand how the digital world is transforming the way we live and do business*—and a specific role: to lead an international research program known as the Information Systems Dynamics (ISD) program.

Today, the ISD program involves:

- **29 international projects**, conducted by some **fifty** research **laboratories** in the United States, Europe, China and Japan;
- Projects based on **field studies** and addressing **themes of strategic importance** for the performance of the digital enterprise.

A.1 The Genesis of the ISD Program

A.1.1 Preamble

On **June 13, 2008,** Senator Pierre Laffitte, President of the Sophia Antipolis Foundation, and Didier Lambert, President of CIGREF, signed an agreement at Opio ratifying the **creation of a research foundation**—the CIGREF Foundation— with the goal of gaining a *"better understanding of how the digital world is transforming the way we live and do business"*.

© Springer International Publishing Switzerland 2016
A. Bounfour, *Digital Futures, Digital Transformation*,
Progress in IS, DOI 10.1007/978-3-319-23279-9

A.1.2 Results of the ISD Program Prototyping Phase (Pre-2010)

A.1.2.1 The ASE-ISD Seminar

The ASE-ISD Seminar was a watershed moment in the history of the ISD program, generating lively interest among its participants—from the business world and academia—in working together. The seminar highlighted 8 key findings:

- It makes more sense to study what IT has done for **society** rather than for business: at the end of the day, the target is the user rather, not the employee;
- Great care must be taken, in this program, to **retrace history** not for historians, but **for citizens**;
- How does **history relate to the future**? Basically, history repeats itself, so we need to be aware of it. There's a word for this: experience. And experience is there to be learned from;
- Attention must be paid to the **cultural approach**: the company culture has a direct impact on IT usage;
- From a training/education perspective, there is a divide between the academic world and the business world: our values are a long way from the "collaborative mode" that represents the future in business. A manager must be more than just a hierarchical superior: he or she must be able to get employees to work together. This should lead us to **rethink our social practices**;
- The **choice of the people** to implement this program will play a critical role in its success;
- It is important that the next debates are seen to be more representative in terms of gender and international scope;
- If the deliverable were a publication, it would take the form of **3 books**: The **History of the Digital World**, The **Geography of the Digital World**, and a **Users' Guide to the Digital World**.

A.1.2.2 Prospective Work

This review of forward-looking studies set out to establish the ISD program's **legitimacy in prospective analysis**. The review concluded that:

- The ISD agenda supplements the agenda of other international programs (EPO, FISTERA, Futuris, research by CAS, MIT, the OECD, etc.);
- The **organizational dimension is key**, encompassing a plurality of spaces (including from the viewpoint of large corporations);

- This should be analyzed in parallel with the **social and ethical dimension**, taking account of technological advances in the broad sense (notably in NBIC: Nanotechnology, Biotechnology, Information technology and Cognitive science);
- **Macro scenarios** and **regional specificities** can play an important role in types of use and typical configurations.

A.1.2.3 Historical Work

Three historical studies were subsequently conducted and published **in a special issue of the journal** *Entreprises et Histoire*, on December 16, 2010:

- A paper on the **economic and institutional environment**, with two case studies (Axa and Total) by Alain Beltran, Research Director at the CNRS, and Pascal Griset, Professor at Paris-Sorbonne University;
- A paper by Alexandre Giandou, ISH, CNRS, drawing on the CIGREF archives to sketch out CIGREF's vision, as an institution, of the **dynamics of IT usage**, and of its own historic role in this development;
- A paper on the **research agenda**: a review of academic research on IT (1978–2008) (more than 2200 publications) by Prof. Bernard Fallery, Sylvie Desq, Aurélie Girard, and Florence Rodhain from CREGOR, University of Montpellier II.

A.1.2.4 Work on Generational Issues

In 2009–2010, a number of workshops were held at CIGREF to discuss Generation Y.

- The workshops established the conceptual framework for an in-depth analysis of generational behaviors with regard to the use of IT;
- This analysis will be extended, under the ISD program, to other functional and geographic contexts (cf. the Telecom Business School project in Wave A);
- Correlation with other studies (such as those by CEFRIO) on other generations is in progress.

A.1.2.5 The First ISD Conference: Initial Inventory of IT Use—Guest Country: Japan

The 1st ISD conference successfully:

- Demonstrated the program's capacity for implementation;
- Affirmed ISD's institutional status as an international research program, and currently the only one of its kind;

- Marked the effective start date for the research work;
- Established an initial state of play on IT use, with a guest country: Japan.

A.2   CIGREF Foundation Governance

The Foundation and (at the operational level) the ISD research program are managed through several Committees: **the Steering Committee, the Scientific Committee, the Strategic Orientation Committee, the History Committee and the Organizing Committee**.

A.2.1   The Steering Committee

The Steering Committee, chaired by CIGREF's President, **defines the Foundation's main lines of action**, as well as its **activity programs and communication policy**. It defines and validates the strategic orientations, and has the last word on all decisions about the running of the program. It coopts qualified personalities, who are called upon to sit on the Foundation's various committees.

A.2.2   The Scientific Committee (SC)

The Scientific Committee is an essential component in the governance of the ISD research program. Its members, nominated by the program's General Rapporteur, are tasked with **bringing their expertise to bear on the scientific dimension** of the program's activities. They also verify that projects are **innovative and worthy of research**.

The Scientific Committee is coordinated by the program's General Rapporteur, Professor Ahmed Bounfour, who holds the European Chair in Intellectual Capital Management at University Paris-Sud.

The membership of the Scientific Committee is as follows:

- Jean-Eric AUBERT, World Bank Institute
- Prof. Surinder K. BATRA, Institute of Management Technology Ghaziabad
- Prof. Michel BEAUDOUIN-LAFON, University Paris-Sud
- Prof. Pierre-Jean BENGHOZI, Ecole Polytechnique
- Prof. Marcos CAVALCANTI, Federal University of Rio de Janeiro
- Prof. Leif EDVINSSON, University of Lund
- Patrick FRIDENSON, EHESS, Centre de Recherches Historiques
- Dominique GUELLEC, OECD
- Prof. Tom HOUSEL, Naval Postgraduate School of Management
- Prof. Junichi IIJIMA, Tokyo Institute of Technology

- Prof. Moez LIMAYEM, University of South Florida
- Prof. Rik MAES, University of Amsterdam
- Prof. M. Lynne MARKUS, Bentley University
- Prof. Peter MEUSBURGER, University of Heidelberg
- Prof. Ian MILES, University of Manchester
- Prof. Yves PIGNEUR, University of Lausanne
- Prof. Frantz ROWE, University of Nantes
- Gérald SANTUCCI, European Commission
- Prof. Pirjo STAHLE, Aalto University School of Engineering
- Prof. Eric TSUI, Hong Kong Polytechnic University.

A.2.3 The Strategic Orientation Committee (SOC)

The mission of the Strategic Orientation Committee—chaired by Alain Pouyat (Executive VP, Information Systems and New Technologies, Bouygues Group), member of the *Académie des Technologies*—is to **act as the vital relay between the ISD program and the business world**. Its members represent the interest shown by top management teams in how information systems and their underlying technologies are used. The SOC suggests **lines of enquiry** to the Foundation's Steering Committee **that it sees as useful** to the pursuit of the program.

A.2.4 The History Committee

The History Committee, chaired by former CIGREF President Didier Lambert, deals specifically with the historical dimension of the program, with a view to learning from the collective memory of IT use. It works in close collaboration with the other Committees, in particular the Scientific Committee, for the methodological aspects and the programming of research activities.

A.2.5 The Organizing Committee

The Organizing Committee, chaired by Jean-François Pépin (Secretary General of the CIGREF Foundation), is in charge of the day-to-day operational management of the program. It reports to the Steering Committee on the effective implementation of the program, in line with the decisions made at Steering Committee level.

Regular **meetings and dinner-debates** were organized between these different bodies in 2010–2014 in order to **promote joint thinking** about progress on the ISD program and the associated issues.

The members of the CIGREF Foundation's Scientific Committee, notably Professors Pierre-Jean Benghozi, Pirjo Stahle, Peter Meusburger, Ian Miles and

Frantz Rowe, as well as Jean-Eric Aubert, were interviewed about ISD, their involvement in the program, and the specific topics addressed. They were also invited to speak about their areas of research (forward-looking studies on the business of the future, new digital business models, the geography of knowledge, etc.). The interviews with the members of the SC can be found on the CIGREF Foundation website.

A.3   ISD Program Rollout Phase (2010–2012)

A.3.1   The Research Work: Wave A Projects

On **March 15, 2010**, the CIGREF Foundation issued its **first call for research projects**. Numerous applications were received, from laboratories in France and around the world. Following an assessment process conducted by an international scientific committee, nine projects were selected:

A.3.1.1   Research Work Packages Covered

- WP 1: Business models
- WP 6: Human resources, work organization and collective intelligence
- WP 9: Emerging practices
- WP 13: Blank call for projects

A.3.1.2   Calendar

- Start of projects: September 2011
- Interim results: April 2011
- Final results: June 2011

A.3.1.3   Teams Involved and Projects

9 projects were selected in this call for projects:

- HEC—Mines: *Coproduction de valeur et systèmes d'information.*
  Scientific lead: Marie-Hélène Delmond
- University of Southern California: *Towards a unified framework for business modeling in the evolving digital space: identifying the co-creation of value with customers, complementors, competitors and community.*
  Scientific leads: Omar El Sawy and Francis Perreira

- University of Montpellier II, Univ. of Twente and LEST, Univ. of Aix-Marseille: *Usages des outils d'intelligence collective: analyser le rôle de la structure organisationnelle.*
  Scientific lead: Bernard Fallery
- Bordeaux School of Management: *Organisational and IS configurations for exploration and exploitation trade-off: the case of a multinational company.*
  Scientific lead: Olivier Dupouet
- Telecom Business School: *Génération Y et pratiques de management des projets SI.*
  Scientific lead: Chantal Morley
- University of Technology, Troyes—ESCEM: *Définir et évaluer une nouvelle méthodologie s'appuyant sur des technologies innovantes pour étudier des pratiques émergentes dans les activités professionnelles.*
  Scientific leads: Eddie Soulier and François Silva
- Paris Dauphine University—IMRI and M-Lab, Ecole de Management de Normandie: *L'impact du Web 2.0 sur les organisations.*
  Scientific lead: Sébastien Tran
- Hanyang University, Korea: *Use of smart phones for organizational coordination.*
  Scientific lead: Namjae Cho
- Paris Dauphine University: *Pratiques émergentes et systèmes d'information : une perspective multi-niveaux.*
  Scientific lead: François-Xavier de Vaujany.

A.3.1.4 Seminars with the Research Teams

Throughout these projects, the teams met up at CIGREF for **research seminars**, to discuss their research goals, their interim results, and their final analysis. The dialogue was continued on a **collaborative platform** provided for their benefit.

- **Kick-off seminar** (December 2010): pooling the initial research results from the projects selected for Wave A;
- **Interim results presentation seminar** (March 11, 2011): the project teams' videos were put online on the CIGREF Foundation website;
- **Final results presentation seminar** (June 29, 2011). New interviews were conducted with the research teams.

A.3.1.5 Presentation of the Results of Wave A

On **September 1st 2011**, after the **teams had worked on the projects for one year, Professor Bounfour**, General Rapporteur for the ISD program, **presented**

the results of the projects to the CIGREF Foundation's Steering Committee and Strategic Orientation Committee.

This overview identified some of the core components (the first **building blocks**) of the **2020 enterprise**, namely:

- The **transformation of business models by digital**;
- A multiplicity of **value creation spaces**;
- **Mobility** as a **powerful trend** in digital usage;
- Widespread and **diverse collaborative practices**;
- A trend towards the management of a **set of vertical and horizontal links**;
- **Trade-offs** between the **need for security** and the **facilitation of collaboration**.

This perspective also touched on a **new system of production—*Acceluction*.**

Acceluction is a mode of production characterized by both the **expansion of the domain of value production** to multiple spaces and the **instantaneity of exchanges** (transactional or organic). Both of these axes are underscored by the ongoing digital acceleration.

Acceluction, therefore, is a **system defined by the accelerated production of links**. These links may be transactional, i.e. governed by market logic, or organic (governed to varying degrees by recognition).

A.3.1.6   Dissemination of the Results of Wave A

The Foundation Workshops

To disseminate the results of the ISD research program to the business world, the CIGREF Foundation set up a series of encounters entitled the "Foundation Workshops" (*Ateliers de La Fondation*). These Workshops are **aimed at practitioners**, to enable them to engage in **dialogue and reflection with the research teams** involved in Wave A of the ISD program. These encounters are designed in a spirit of **cross-fertilization** between practitioners and researchers.

The first workshop, in March 2012, centered around **"Business models in the digital era: towards a multiplicity of value creation spaces"**. Research was presented by HEC-MINES on **information systems and the coproduction of value**, and by the University of Southern California on **digital business models**.

A second workshop on **new modes of work organization in a digital context** was held on September 21, 2012, under the chairmanship of Professor Bounfour, General Rapporteur of the ISD international research program. The speakers were:

- **Prof. Namjae CHO,** Hanyang University (South Korea), "**Use of smart phones for organizational coordination**" (in English);
- **Prof. Bernard FALLERY**, University of Montpellier 2, "*Usages des outils d'intelligence collective*: *le rôle de la structure organisationnelle*";

- **Sébastien TRAN**, Ecole de Management de Normandie, "*L'impact du web 2.0 sur les organisations*";
- **Prof. Chantal MORLEY**, Telecom Business School, "*Génération Y et pratiques de management des projets SI*".

Academic series : *SpringerBriefs in Digital Spaces* and *Espaces numériques*

Recognition of the ISD program is furthered by the **publication of books and papers**. To this end, CIGREF has established partnerships with prestigious **international publishers**. An agreement has been signed with **Springer Verlag**, one of the world's foremost scientific publishers, to create a new SpringerBriefs series entitled "**Digital Spaces**", partnered by CIGREF, to help **disseminate the results of the program in English**.

The results of the program are also **published in French in book format** in Springer's "*Espaces numériques*" series.

The *Essentials* series

The ISD program's reputation also depends on **digital dissemination** and the organization of **collective debate around its results**. The publication of the "CIGREF Foundation Essentials" offers readers a summary of the work delivered by the researchers.

To date, the Essentials of all Wave A projects are available from the CIGREF website in the form of **enriched e-books, in French and English**.

A.3.2 The Research Work: Wave B Projects

Building on the perspectives provided by this initial overview, a **second wave of projects was launched in 2011**. With "Wave B", the ISD program confirmed its **international reach**, as the 10 projects selected—covering new social and ethical values, open innovation, and knowledge flows—were conducted by leading institutions such as University of South California or Tsingua in China and Meiji university in Japan.

The selected teams were brought together for a seminar at CIGREF on November 14, 2011. Their interviews are available at the Foundation website.

A.3.2.1 Research Work Packages Covered by Wave B

- WP 2: Social and ethical values
- WP 3: Open innovation

- WP 5: Knowledge flows
- WP 6B: Future Enterprise Design 2020
- WP 13: Blank call for projects

A.3.2.2   Wave B Calendar

- Announcement of selected projects: April 5, 2011
- Start of projects: September 2011
- Interim results: March 2012
- Final results: September 2012

A.3.2.3   Teams Involved and Projects

- Greenwich University (UK), EHESS (Ecole des Hautes Etudes en Sciences Sociales, France): *Tester l'hypothèse de la "fin de la vie privée" dans la communication assistée par ordinateur: une approche par la modélisation multi-agent.*
  Scientific leads: Paola Tubaro and Antonio Casilli
- Montfort University (UK): *Identification and governance of emerging ethical issues in information systems.*
  Scientific lead: Bernd Carsten Stahle
- University of Aachen (Germany) in partnership with the University of Tsinghua (China): *Innovating in a learning community.*
  Scientific leads: Kai Reimers and Guo Xunhua
- National University of Sun Yat-sen: *How information technologies affect the knowledge ecology and the adoption of open innovation: a multinational study.*
  Scientific lead: Ting Peng Liang
- Groupe Sup de Co Montpellier Business School: *International knowledge markets effects.*
  Scientific lead: Benbya Hind
- Brunel University (UK): *Globally distributed innovation and co-creation of value: cases of UK-China collaborations.*
  Scientific lead: Pamela Abbott
- University of Southern California (USA): *A framework for assessing the effective use of social media tools in the enterprise to enhance innovation, collective problem solving, knowledge sharing and management of virtual teams.*
  Scientific lead: Ann Majchrzak
- Meiji University: *An East Asian perspective on the developing ethical and social values of digital object usage.*
  Scientific lead: Kiyoshi Murata

- EM Strasbourg Business School: *Observatoire des entreprenants en Système d'Information (SI).*
  Scientific lead: Gaëtan Mourmant

A.3.2.4 Presentation of Initial Project Results for Wave B

Professor Bounfour presented the first results from these projects to the Steering Committee and the Strategic Orientation Committee at a presentation seminar held at Euro Disney on September 5, 2012.

The Expansion of Value Creation Spaces is Confirmed, Notably
by the "Open Innovation" and "Knowledge Flows" Projects

- **Original innovation models between persons with no prior links** can be organized into ad hoc models in a digital platform framework;
- The search for new market- or community-based **incentive mechanisms** attests to the widely expected impact of digital on value creation;
- Monetary incentives are an important but not necessarily predominant component of exchange and transaction processes in digital spaces.

The Emergence of New Social and Ethical Issues

- The ubiquity of digital, with radical changes in the frontiers between activities and organizations, raises **new ethical issues**, while the corresponding codes of practice are still **little more than good intentions**;
- The **ethical risk** is clear, but there is no risk of digital uses being rejected;
- Contrary to the received wisdom, the **exposure of individuals and of personal privacy** is not a uniform phenomenon, and it is neither definitive nor irreversible. **Individuals adjust** to specific situations (for example, in response to the behavior of the owners of social media platforms).

Mapping Out a Specific Space: Entrepreneurial Space

- The **frontier between transactional space and community space** is shown to be **porous**;
- In or around 2020, the digital revolution will bring about fundamental transformations that are already driving enterprises to align with three types of trend:
  - IT spaces of **intrapreneurial freedom**;
  - IT Spaces of **entrepreneurial freedom**;
  - The need for **IT entrepreneurship**.

Overall, the projects confirm the relevance of steering the **design of the 2020 enterprise** towards the definition of a **multiplicity of spaces** and their articulation by **mobilizing digital resources** in accordance with **ad hoc modes of governance**.

A.3.2.5   Dissemination of the Results of Wave B

The *Essentials* Series

To date, the Essentials of all Wave B projects are available from the CIGREF website in the form of **e-books, in French and English**.

A.3.3   The Research Work: Wave C Projects

Following the Foundation's latest call for projects, support was provided for **11 international research projects**, which were conducted from January 2013 to January 2014.

A.3.3.1   Research Work Packages Covered by Wave C

- WP 6B: Future Enterprise Design 2020
- WP 7: Technological convergence, digital use, and digital materiality
- WP 8: Economics of digitality
- WP 10: Intellectual property rights and regulation
- WP 11: Digital maturity, performance and standards
- WP 12: Managing digitals spaces and functions in 2020

A.3.3.2   Teams Involved and Projects

- *Centre d'appel 2020*, Centre d'Etudes de l'Emploi
- *Designing 21st century organizations for generativity: An organizational genetics approach*, Temple University (USA)
- *Sociotechnical designs for 2020 R&D enterprises: accelerating innovation by emergently leveraging global distributed knowledge, human capital, and digital assets*, University of Washington, University of Southern California
- *Learning from M2M business models: implications for the business enterprise 2020*, University of Southern California
- *Changement et adaptation individuelle aux innovations organisationnelles et aux technologies émergentes: adoption, meilleures pratiques et performance*, Paris Dauphine University, EM Strasbourg School of Management

- *Pour un modèle de maturité des espaces numériques pour les PME : passer d'une «envie de technologie web» à une stratégie de technologie web*, University of Aix-Marseille
- *Relationship between information technology governance configuration and organizational performance*, George Mason University
- *Intellectual property law and freedom: between the national and the international*, University of Cambridge
- *Re-defining the space for company-community interaction: How can firms leverage the innovative potential of open source software production model?* University of Greenwich
- *Towards a maturity model for the assessment of ideation processes in crowdsourcing projects*, University of Nebraska (Center for Collaboration Science), TELECOM ParisTech
- *Corporate standardisation management in the ICT Sector*, RWTH Aachen University

A.3.3.3   Seminars with the Research Teams

Throughout the duration of these projects, the teams met up via videoconference at **research seminars** to discuss their research goals, their interim results, and their final analysis. The dialogue was continued on a **collaborative platform** provided for their benefit.

- **Kick-off seminar** (December 2013)
- **Interim results presentation seminar** (April 5, 2013)
- **Final results presentation seminar** (January 23, 2014).

A.3.3.4   Presentation of Initial Project Results for Wave C

The aim of these projects was to **round out the design of the 2020 enterprise** by examining the following issues in greater depth:

- Accelerating innovation by drawing on distributed knowledge, human capital and digital assets;
- Adopting a multidisciplinary (management and genetics) approach to describe the generativity of tomorrow's enterprise;
- Describing change and individual adaptation to organizational innovations and emerging technologies;
- Characterizing relations between IT configuration and organizational performance;

- Defining maturity models in the evaluation of value creation by crowdsourcing;
- Improving innovation through company-community interactions;
- Questions around standardization in the new digital world were also addressed.

A.3.3.5   Dissemination of the Results of Wave C

The *Essentials* Series

The Essentials of the Wave C projects C are available from the CIGREF website, in French and English.

A.3.4   International Events

In 2010, the CIGREF Foundation organized the 2nd edition of its major public conference dedicated to digital innovation and China. Since then, the Foundation has also intervened in a number of **international events** to present the **results** of the ISD program.

A.3.4.1   2nd ISD Conference (September 23, 2010)

The 2nd conference focused on **digital innovation at the service of business transformation**, and the guest country was China. Conference takeaways included:

- Facing up to the **foreseeable impact of the rise of a great global power** (China):
  - Parity with the USA by 2020?
  - A clear digital strategy with national champions (cf. contribution from Taobao)
  - Similarity of issues around use
- Recognition that IS **uses are evolving** towards the digital enterprise
- **International comparison** of organizational innovations and business models (Canada, USA, Taiwan, etc.)
- Deliverables available at the CIGREF Foundation website:
  - 11 conference highlights (videos + presentations)
  - A 5-chapter summary with a "key-points" reference document.

A.3.4.2 CIO Magazine Conference (November 2011)

In November 2011, CIGREF and the CIGREF Foundation partnered, along with CIO Magazine, a conference on "**Digital Innovation in Africa**" held at the Bibiliothèque Nationale in Paris. In the opening speech, Bruno Ménard outlined the ISD program. Professor Bounfour presented the program's initial results, and in particular the concept of acceluction.

A.3.4.3 ICIS Shanghai (December 2011)

The International Conference on Information Systems (ICIS), one of the most prestigious gatherings of information system practitioners and academics, was held in Shanghai (China) from December 4–7, 2011 on the theme: **East Meets West: Connectivity and Collaboration through Effective Information Systems**.

A specific session was devoted to the CIGREF Foundation's ISD research program, as part of the **CIO Symposium**, on Tuesday December 6 from 12:00 to 14:00, with speeches from:

- **Bruno Ménard**, who gave the keynote speech (by video);
- **Prof. A. Bounfour**, who présented the ISD program and the first results of Wave A;
- **Prof. Ting Peng Liang** (National University of Sun Yat-Sen)**, Dr. Pamela Abbot** (Brunel University) **and Prof. Guo Xunhua** (Tsinghua University)—**3 research teams** from the projects supported by ISD as part of Wave B;
- **Profs. Moez Limayem, Frantz Rowe** (members of the Scientific Committee) and **Omar El Sawy** (director of a Wave A research project) at round-table discussion.

A.3.4.4 Global Forum, Stockholm (November 2012)

The theme for this year's Global Forum conference, which convenes a gathering of international decision-makers every year, was: "*Sharing a Connected Digital Future: Visions, Challenges, Opportunities for Organizations and People in a Smart World*". Organized by Items, in partnership with Vinnova, Sweden's national agency for innovation systems, it was held in Stockholm on November 12 and 13, 2012.

The CIGREF Foundation's ISD research program was the guest contributor at the opening session on November 13, devoted to innovation and open innovation policies.

A.3.4.5 Intellectual Capital 9 (IC9): Emerging Worlds, Growing Intangibles, World Bank, Paris, (June 6 and 7, 2013)

One of the sessions at IC9 was devoted to digital transformation, business models and data use. It brought together researchers from different countries, all members of the ISD international research program, including Gérald Santucci (from the European Commission), Professor Omar El Sawy (University of Southern California), Professor Namjae Cho (Hengyang University, Korea) and Dr. Pamela Abbott from Brunel University (UK). The session was opened by Pascal Buffard, who came to present the design of the 2020 enterprise.

A.3.4.6 Meeting with the DG Connect at the European Commission (September 2013)

On September 3, 2013, Pascal Buffard and Prof. Bounfour went to the European Commission to present the results of the ISD program to the director general of the DG Connect, Mr. Robert Madelin and his team, for comparison with the European research agenda.

A.3.4.7 EuroCIO, Brussels, (November 27 and 28, 2013)

The CIGREF Foundation headed *Workshop* #1 *"How to design the 2020 enterprise"*. Chaired by CIGREF board member Konstantinos Voyiatzis, the Global CIO of Nexans Group, the workshop built on the results of the ISD international research program, aimed at furnishing the conceptual building bricks for the design of the 2020 enterprise and associated digital resources. A series of questions were addressed, including:

- What will the value spaces be for the 2020 enterprise?
- Which scenarios could be envisaged (markets, community platforms, hybrid forms, etc.) for your company in 2020?
- How, in concrete terms, will the new generation of organizational models impact IT governance?
- How will digital resources contribute to and/or spearhead such a transformation?
- What are the implications for the design of the new IT/IS function?

A.3.4.8 International Collaborations and Affiliated Projects

Finally, the CIGREF Foundation has developed a series of **relationships with organizations in other countries** in areas relating to the ISD program:

- AUSIM, **Morocco** (formal agreement)

  - Agreement signed on June 18, 2010
  - Research topics: SMEs

- CEFRIO, **Canada** (formal agreement)

  - Research collaboration agreement signed on June 23, 2009
  - Collaboration topics: Generation C, innovation index

- **Taiwan**

  - A deliverable on Product Lifecycle Management (PLM).

Annexe B
ISD Projects Presented in Figs. 3.1–3.5

Acronym	Research institution	Names of researchers involved	Name of the project
Business models and ecosystems of innovation			
CV and SI	HEC-Mines	Marie-Hélène Delmond, Alain Keravel, Alain Busson, Robert Mahl, Fabien Coelho	Coproduction de valeur et systèmes d'information
UniFoBM	University of South California	Omar El Sawy, Francis Pereira	Towards a unified framework for business modelling in the evolving digital space: identifying the co-creation of value with customers, complementors, competitors and community
MN-PME2012	Université de Toulouse, Université Toulouse 1 Capitole, et Aix-Marseille Université	Bénédicte Aldebert, Marie-Christine Monnoyer	Pour un modèle de maturité des espaces numériques pour les PME: passer d'une «envie de technologie web» à une stratégie de technologie web
PLM	National Sun Yat-sen University, National Chung Cheng University, I-Shou University	Chin-Fu Ho, Wei-Hsi Hung, Kao-Hui Kung	PLM usage behavior and technology adaptation
Mobility			
SMC	Hanyang University (Korea)	Namjae Cho	Use of smart phones for organizational coordination

(continued)

© Springer International Publishing Switzerland 2016
A. Bounfour, *Digital Futures, Digital Transformation*,
Progress in IS, DOI 10.1007/978-3-319-23279-9

Acronym	Research institution	Names of researchers involved	Name of the project
Work, coordination and the generational issue			
PMY	Télécom Ecole de Management	Chantal Morley, Said Assar, Marie Bia Figueiredo, Imed Boughzala, Marité Milon, Thierno Tounkara	Génération Y et pratiques de management des projets SI
USE.ORG	Université Montpellier II	Bernard Fallery, Roxana Oloeanu-Taddei, Tanya Bondarouk, Huub Ruel, Ewan Oiry, Amandine pascal, Robert Tchobanian, Yael Guillon	Usages des outils d'intelligence collective: analyser le rôle de la structure ORGanisationnelle
Web 2.0	Université Paris Dauphine—IMRI and M-Lab Ecole de Management de Normandie	Sébastien Tran, Sonia Cheffi, Sébastien Damart, Albert David, Amir Hasnaoui, Nicolas Monomakhoff, Luisa Zibara	L'impact du Web 2.0 sur les organisations
CA2020	Centre d'Etudes de l'Emploi	Nathalie Greenan	Centre d'appel 2020
Generation C	CEFRIO	Réjean Roy, Philippe Aubé, Catherine Lamy	La génération C—Les 12–24 ans: moteurs de transformation des organisations
Emerging uses, individual adaptation			
METEPE	Université de Technologie de Troyes	Eddie Soulier, François Sylva, Yves Boisselier	Définir et évaluer une nouvelle méthodologie s'appuyant sur des technologies innovantes pour étudier des pratiques émergentes dans les activités professionnelles
CAITI	Université Paris Dauphine, Strasbourg Ecole de Management	Christophe Elie-dit-Cosaque, Jessie Pallud	Changement et adaptation individuelle aux innovations organisationnelles et aux technologies émergentes: adoption, meilleures pratiques et performance

(continued)

Acronym	Research institution	Names of researchers involved	Name of the project
MLA	Université Paris Dauphine	François-Xavier de Vaujany, Sabrine Carton, Carine Dominguez, Emmanuelle Waast	Pratiques émergentes et systèmes d'information: une perspective multi-niveaux
Internal innovation			
ORISCO	BEM Bordeaux School of Management	Olivier Dupouet, C. Lakshman, T. Bouzdine-Chameeva	ORganisational and IS COnfigurations for exploration and exploitation trade-off: the case of a multinational company
IKME	Groupe Sup de Co Montpellier Business School, MIT	Benbya Hind, Nassim Belbaly, Marc Robert, Marshall Van Alstyne, Bill McKelvey	Internal knowledge markets effects
SMI	University of Southern California	Ann Majchrzak, Elisabeth Fife, Francis Pereira	A framework for understanding the use of social media tools in the enterprise to enhance innovation: a cross cultural approach
Open innovation and Knowledge flow			
ILC	Aachen University, Tsinghua University	Kai Reimers, Chen Guoqing, Guo Xunhua, Li Mingzhi, Bin Xie	Innovating in a learning community
KEOIA	National University of Sun Yat-sen	Ting Peng Liang, LG Pee, Dai Seno, Huang Lihua	How information technologies affect the knowledge ecology and their adoption of open innovation: a multinational study
SCCI	University of Greenwich	Riccardo DE VITA, Guido Conaldi	Re-defining the space for companies-communities interaction: how can firms leverage the innovative potential of open source software production model?
GLOBVAL	Brunel University	Pamela Abbott, Yingqin Zheng	Globally distributed innovation and co-creation of value: cases of UK-China collaborations

(continued)

Acronym	Research institution	Names of researchers involved	Name of the project
CINAM	University of Nebraska, TELECOM ParisTech	Gert-Jan de VREEDE, Imed BOUGHZALA	Towards a maturity model for the assessment of ideation processes in crowdsourcing projects
Ethics of digital usage and privacy			
THEOP	University of Greenwich, London; EHESS, France	Paola Tubaro, Antonio Casilli	Tester l'hypothèse de la "fin de la vie privée" dans la communication assistée par ordinateur: une approche par la modélisation multi-agent
IDEGOV	De Monfort University, Notre Dame de Namur	Bernd Carsten Stahle, Philippe Goujon, Laurence Masclet, Kutoma Wakunuma, Yingqin Zheng	Identification and governance of emerging ethical issues in information systems
DESVALDO	Meiji University	Kiyoshi Murata, Yutaka Takahashi, Thomas Lennerfors, Ryoko Asai, Andrew Adams, Yokho Orito	An East Asian perspective on the developing ethical and social values of digital object usage
Regulation, norms and standards			
IPFL	University of Cambridge	GE Chen	Intellectual property law and freedom: between the national and the International
COSMICS	RWTH Aachen University	Kai Jakobs	COrporate Standardisation Management in the ICt Sector
Economic performance			
ITGovOP	George Mason University	Amitava Dutta, Nirup G. Menon	Relationship between information technology governance configuration and organisational performance
Data			
M2Mod	University of Southern California	Francis Pereira, Omar El Sawy, Ron Ploof	Learning from M2M business models: implications for the business enterprise 2020

(continued)

Acronym	Research institution	Names of researchers involved	Name of the project
Design of the 2020 enterprise			
ODESI	Ecole de management de Strasbourg	Gaëtan Mourmant, Michel Kalika, Guittard Claude, Mlaiki Alya, Pallud Jessie, Schenk Eric, Tiphaine Dalmas	Observatoire des entreprenants en Système d'Information (SI)
Generative M	Temple University (USA)	Youngjin Yoo, Rob Kulathinal, Sunil Wattal	Designing 21st century organizations for generativity: an organizational genetics approach
TMD	University of Washington, University of Southern California	Paul Collins, Ann Majchrzak,	Sociotechnical designs for 2020 R&D enterprises: accelerating innovation by emergently leveraging global distributed knowledge, human capital, and digital assets

Annexe C
SpringerBriefs in Digital Spaces Series

SpringerBriefs in Digital Spaces http://www.springer.com/series/10461

SpringerBriefs in Digital Spaces

Series Ed.: A. Bounfour

The Springer Briefs Series on Digital Spaces is a joint initiative taken recently by CIGREF, Springer and Professor Ahmed Bounfour, as Editor. The series aim at disseminating internationally the results of the ISD international research programme, initiated by CIGREF in 2009, as well as to providing a unique platform for an international dialogue among scholars, policy and business audiences, on the emerging use of digital artifacts and systems.

ISD is an international research programme of public interest that aims to evaluate the societal and managerial challenges in the long-term usage of IS (1970-2020), by mobilising the best expertise available at the international level. Its purpose is to analyse the different dimensions of use of IS and their interactions.

ISD considers that the issue of use of information systems now goes beyond the scope of managerial action: it embraces the whole society. From this central perspective, the programme's objectives are of different types:

• An overall objective: to understand the many facets involved in the dynamic use of information systems over a long period, especially by focusing on emerging factors in different geographical and business contexts;

• A specific objective: to provide the stakeholders (large companies, IT providers, government, academics, media) with the analytical grids that will enable to understand the strategic issues arising from the changes under way

The programme considers that the future of enterprises—and the design of their future IS—will be determined by the interaction between developments in socio-ethical, strategic, technological, regulatory and organisational trends. It is by considering these five perspectives, interactively and systemically, that we can grasp the reality of the driving forces affecting future companies and their information systems.

Since its launch in 2009, the programme is already supporting 20 projects conducted by international teams (From Europe, North America and Asia).

Recently published:

H. Benbya
Exploring the Design and Effects of Internal Knowledge Markets

P. Abbott, Y. Zheng, R. Du
Collaboration, Learning and Innovation Across Outsourced Services Value Networks
Software Services Outsourcing in China

Springer books available as

 Printed book

Available from springer.com/shop

 eBook

Available from your library or
▸ springer.com/shop

 MyCopy

Printed eBook for just
▸ € | $ 24.99
▸ springer.com/mycopy

© Springer International Publishing Switzerland 2016
A. Bounfour, *Digital Futures, Digital Transformation*,
Progress in IS, DOI 10.1007/978-3-319-23279-9

References

Abbot, P., & Zheng, Y. (2012). *Globally distributed innovation and co-creation of value: Cases of UK-China collaborations*. Paris: CIGREF Foundation.

Abbott, P., Zheng, Y., & Du, R. (2014). *Collaboration, learning and innovation across outsourced services value networks, software services outsourcing in China. Springerbriefs in digital spaces*. Heidelberg: Springer.

Adams, A., & Muarata, K. (2013). *An East Asian perspective on the developing ethical and social values of digital object usage*. Paris: CIGREF Foundation.

Alchian, A. & Demsetz, H. (1972). Production information, costs and economic organization. *American Economic Review, 66*, 777–795.

Aldebert, B., Bertrand, D., Monnoyer, M. C., & Seck, A. M. (2014). *Les systèmes d'information inter-organisationnels et la création de valeur: résultats d'une étude menée auprès de deux grandes entreprises françaises Eurocopter et Cegelec-Vinci et leurs fournisseurs*. Paris: Fondation CIGREF.

Atkinson, R. D. (2004). *The past and future of America's economy: Long waves of innovation that power cycles of growth*. Cheltenham: Edward Elgar.

Aubert, J. E. & Wermelinger, M. (2015). Securing livelihoods for all: A OECD Development Centre report, presentation at IC 11. In *The World Conference on Intellectual Capital for Communities*. Available at: www.chaireonintellectualcapital.u-psud.fr

Baudrillard, J. (1972). *Pour une critique de l'économie politique du signe*. Paris: Editions Gallimard.

Baudrillard, J. (1981). *Simulacres et simulation*. Paris: Editions Gallilée.

Bauman, Z. (2000). *Liquid modernity*. Malden, MA: Blackwell Publishing.

Bauman, Z. (2007). *Le présent liquide. Peurs sociales et obsession sécuritaire*. Paris: Le Seuil.

Bauman, Z., & Lyon, D. (2013). *Liquid surveillance*. Malden, MA: Polity Press.

Baxter, R., & Matear, S. (2004). Measuring intangible value in business-to-business buyer-seller relationships: an intellectual capital perspective. *Industrial Marketing Management,33*, 491–500.

Beltran, A. (2010). Arrivée de l'informatique et organisation des entreprises françaises (fin des années 1960- début des années 1980). *Entreprises et Histoire*, n°60, 122–137.

Benbya, H. (2013). *Exploring the design and effects of internal knowledge markets*. Paris: CIGREF Foundation.

Bennis, W. (2013). Leadership in a digital world: Embracing transparency and adaptive capacity. *MIS Quarterly,37*(2), 635–636.

Berner, M., Graupner, E., & Maedche, A. (2014). The information panopticon in the big data era. *Organization Design, JOD,3*(1), 14–19.

Bharadwaj, A., El Sawy, O. A., Pavlou, P. A., & Venkatraman, N. (2013). Digital business strategy: Toward a next generation of insights. *MIS Quarterly,37*(2), 471–482.

Bounfour, A. (1999). Is outsourcing of intangibles a real source of competitive advantage? *Journal of Applied Quality Management,2*(2), 127–151.

© Springer International Publishing Switzerland 2016

A. Bounfour, *Digital Futures, Digital Transformation,*
Progress in IS, DOI 10.1007/978-3-319-23279-9

Bounfour, A. (2005). Modeling intangibles: Transaction regime versus Community regigme. In A. Bounfour & L. Edvinsson (Eds.), *Intellectual capital for communities, nations, regions and cities*. Elsevier Butterworth-Heinemann: Burlington, MA.

Bounfour, A. (2006). *Capital immatériel, connaissance et performance*. Paris: L'Harmattan.

Bounfour, A. (Ed.). (2009). *Organisational capital: Modelling, measuring, contextualising*. London: Routledge.

Bounfour, A. (Ed.). (2010a). *The ISD research agenda*, Paris. CIGREF Foundation. Available at: www.fondation-cigref.org

Bounfour, A. (2010b). Les systèmes d'information: des objets-frontières de la transformation des entreprises. *Entreprises et Histoire*, n°60, 7–16.

Bounfour, A. (2011). *Accelution in action*. CIGREF Foundation: www.fondation-cigref.org

Bounfour, A. (2013). *ISD international research program: An overview of wave B project*. Paris: CIGREF Foundation.

Bounfour, A. (2014). *An overview of wave C projects*. Paris: CIGREF Foundation.

Bounfour, A. Du capitalisme au Communautalisme (forthcoming)

Bounfour, A., & Epinette, G. (2006). *Valeur et performance des SI*. Paris, Dunod: Une nouvelle approche du capital immatériel.

Brynjolfsson, E., Hitt, L., & Kim, H. (2011). Strength in number: How does data-driven decision-aking affect firm performance. In *32nd Conference on Information Systems*, Shangai. Available at: http://aisel.aisnet.org/icis2011/proceedings/economicvalueIS/13/

Bughin, J., & Manyika, J. (2012). *Internet matters, essays in digital transformation*. New York: McKinsey & Company.

Cap Gemini, MIT Sloan Management. (2011). Digital transformation: A roadmap for billion

Carr, N. G. (2003). IT doesn't matter. *Harvard Business Review*, 41–49

Casilli, A., & Tubaro, P. (2012). *Testing the "end of privacy" hypothesis in computer-mediated communication: an agent-based modelling approach*. Paris: CIGREF Foundation.

Castells, M. (2000). *The rise of the networked society, The information age. Economy, society and culture*. Malden, Mass: Blackwell Publishers.

Chandler, A. D. (1992). Organizational Capabilities and the theory of the firm. *Journal of Economic Perspectives,6*(3), 3–28.

Chandler, A. D., & Cortada, J. W. (Eds.). (2000). *A nation transformed by information: How information has shaped the united states from colonial times to the present*. New York: Oxford University Press.

Chesbrough, H. (2006). *Open business models, how to thrive in the new innovation landscape*. Boston: Harvard Business School Press.

Chesbrough, H., Vanhaverbeke, W., & West, J. (2006). *Open innovation, researching a new paradigm*. New York: Oxford University Press.

Cigref. (2013). *Entreprises et culture numérique*. Paris: CIGREF (Preface Pascal Buffard).

Coase, R. H. (1937). The nature of the firm. *Economica,4*(16), 571–586.

Copeland, D. G., & McKenney, J. L. (1988). Airline reservations systems: Lessons from history. *MIS Quarterly*, 353–370

Cortada, J. W. (2004). *The digital hand, Vol. 1: How computers changed the work of american manufacturing, transportation and retail industries*. New York: Oxford University Press.

Cortada, J. W. (2006). *The digital hand, Vol. 2: How computers changed the world of american financial, telecommunications, media, and entertainment industries*. New York: Oxford University Press

Cortada, J. W. (2008). *The digital hand, Vol. 3: How computers changed the work of american public sector industries*. New York: Oxford University Press

Cortada, J. W. (2010). How IT makes industries so important. *Entreprises et Histoire*, n°6°, 29–49.

Crawford, M. B. (2009). *Shop class as soul craft. An inquiry into the value o work*. Penguin Press. Translation in French: *Eloge du carburateur. Essai sur le sens et la valeur du travail*. Paris, La Découverte (2010).

Culnan, M. J., & Swanson, E. B. (1986). Research in management information systems, 1980–1984, points of work and reference. *MIS Quarterly,10*(3), 289–301.

Davenport, T. (2014). *Big data @ work*. Boston, MA: Harvard Business School Publishing.

De Soto, H. (2001). *The mystery of capital: why capitalism triumphs in the west and fails everywhere*. London : Blackswan.

De Vita, R., & Conaldi, G. (2014). *Re-defining the space for companies-communities interaction: How can firms leverage the innovative potential of open source software production model?*. Paris: CIGREF Foundation.

Dehning, B., Richardson, V. J., & Zmund, R. W. (2003). The value relevance of announcements of transformational information technology investments. *MIS Quarterly, 27*, 637–656.

Desq, S., Fallery, B., Reix, R., & Rodhain, F. (2002). 25 ans de recherche en systèmes d'information. *Systèmes d'Information et Management,7*(3), 5–31.

Desq, S., Reix, R., Rodhain, F., & Fallery, B. (2007). La spécificité de la recherche française en systèmes d'information. *Revue française de gestion,176*, 2007.

Drnevich, P. L., & Croson, D. C. (2013). Information technology and business level strategy: Toward an integrated theoretical perspective. *MIS Quarterly,37*(2), 483–509.

Dutta, A., & Menon, N. M. (2013). *IT configurations and organizational performance*. Paris: CIGREF Foundation.

El Sawy, O. A., & Pereira, F. (2011). *Towards a unified framework for business modelling in the evolving digital space*. Paris: CIGREF Foundation.

El Sawy, O. A., Pereira, F., & Ploof, R. (2014). *Rethinking enterprise 2020 for networked abundance in the value-shifting age of the internet of things*. Paris: CIGREF Foundation.

ElieDitCosaque, C., & Pallud, J. (2014). *Changement et adaptation individuelle aux innovations organisationnelles et aux technologies émergentes: adoption, meilleures pratiques et performance*. Paris: Fondation CIGREF.

Eriksen, T. H. (2001). *Tyranny of the moment, fast and slow time in the information age*. London: Pluto Press.

Eurocio. (2013). Workshop "How to design the 2020 enterprise" report from Worskp 1, Brussels, December

European Internet Foundation. (2014). *The digital world in 2030, what place for Europe?* Available at: https://www.eifonline.org/the-digital-world-in-2030.html

European Patent Office. (2007). *Sceanarios for the future how might IP regimes evolves by 2025? What global legitimacy might such regimes have?*. Munich: EPO.

Foucault, M. (1975). *Surveiller et punir*. Paris: Gallimard.

Freeman, C., & Louça, F. (2001). *As times goes by: From the Industrial revolutions to the information revolution*. Oxford: Oxford University Press.

Future Internet Enteprise Systems (FInES). (2012). Research roadmap 2025 Brussels. Available at: http://cordis.europa.eu/fp7/ict/enet/documents/fines-research-roadmap-v30_en.pdf

Galbraith, J. R. (2014). Organizational design, challenges resulting from Big data. *Journal of Organization Design, JOD, 3*(1), 2–13.

Galloway, A. (2012). *The interface effect*. Cambridge, UK: Polity Press.

Ge, C. (2013). *Intellectual property law and freedom: Between the national and the International*. Paris: CIGREF Foundation.

Giandou, A. (2010). Le CIGREF: un club d'entreprises, acteur majeur de l'évolution des systèmes d'information (1070–2010). *Entreprises et Histoire,60*, 63–77.

Gille, L., & Marchandise, J.-F. (2013). *La dynamique internet, prospective 2030, Paris, Commissariat Général à la Stratégie et à la Prospective*. Available at: http://archives.strategie.gouv.fr/cas/content/etude-dynamique-internet-2030.html

Greenan, N. D., Guillemot, D., & Kocoglu, Y. (2010). Informatisation et changements organisationnels dans les entreprises. *Réseaux*, June–July 2010

Greenan, N., & Mairesse, J. (2006). Les changements organisationnels, l'informatisation des entreprises et le travail des salariés. Un exercice de mesure à partir des données couplées. *Revue Économique*, 57(6, 2), 1137–1175.

Griset, P. (2010). du «temps réel» aux premiers réseaux: une entreprise rêvée, une informatique à l'épreuve du quotidien. *Entreprises et Histoire,60*, 99–121.

Gulati, R., Puranam, P., & Tushman, M. (2012). Meta-organiszation design: Rethinking design in interorganizatuonal and community contexts. *Stratégic Management Journal,33*(6), 571–771.

Haigh, N., Walker, J., Bacq, S., & Kickul, J. (2015). Hybrid organizations, origins, strategies, impacts and implications. *California Management Review,57*(3), 5–12.

Harvey, D. (1989). *The conditions of Post-Modernity.* Wiley-Blackwell.

Harvey, D. (2005). *"Spaces of neoliberalzation: Towards a theory of uneven geographical development" Hettner-lecture 2004 with David Harvey.* Stuttgart: Franz Steiner Verlag.

Harvey, D. (2009). *The urban experience.* Baltimore & London: The Johns Hopkins University Press.

Hassan, R. (2003). *The chronoscopic society, globalization, time and knowledge in the network economy.* New York: Peter Lang Publishing.

Hirschheim, R., & Lacity, M. C. (2000). The myths and realities of information technology insourcing. *Communications of the ACM, 43*(2), 99–107.

Hochereau, F. (2010). Le mouvement d'informatisation d'une grande entreprise: les visions organisantes successives d'un processus d'activité stratégique. *Entreprises et Histoire*, n°60, 138–157.

Honneth, A. (2002). *La lutte pour la reconnaissance.* Paris: Les Editions du Cerf. Collection.

Innovation futures. (2011). *Scenario report: Deliverable D.4.1 (wp4).* Available at: http://www.sustainable-everyday-project.net/innovation-futures/deliverables-and-publications/

Jensen, M. C., & Meckling, W. H. (1976). Theory of the firm: Managerial behavior, agency costs and ownership structure. *Journal of Financial Economics, 3, 4*, 305–360.

Keen, P., & Williams, R. (2013). Value architectures for digital business: Beyond the business model. *MIS Quarterly,37*(2), 643–647.

Krafcik, J. (1988). Triumph of the lean production system. *Sloan Management Review,30*(1), 41–51.

Krugman, P. (1998). Space: The final frontier. *Journal of Economic Perspectives,12*(2), 161–174.

Lee, J. K. (2015). Research framework for AIS grand vision of the bright ICT initiative. *Management Information Systems Quarterly, 39*(2), iiii–xii.

Le Goff, J. (1977). *Pour un autre moyen âge Temps, travail et culture en occident.* Paris: Gallimard.

Lefebvre, H. (2000). *La production de l'espace.* Paris: Editions Anthropos.

Lévy, P. (1998). *Becoming virtual, reality in the digital age.* New York and London: Plenum Trade.

Lucas, H. C., et al. (2013). Impactful research on transformational information technology: An opportunity to information new audiences. *MIS Quarterly,37*(2), 371–378.

Lyon, D. (2007). *Surveillance studies: An Overview.* Malden, MA: Polity Press.

Lyotard, J. F. (1979). *La condition postmoderne, Rapport sur le savoir.* Paris: Les Editions de Minuit.

Malecki, E. J., & Moriset, B. (2008). *The digital economy, business organization, production processes and regional developments.* Abingdon: Routledge.

Malone, T. W., Laubacher, R., & Scott Morton, M. S. (2003). *Inventing the organizations of the 21st Century.* Cambridge, MA: The MIT Press.

March, J. G., & Simon, H. A. (1958). *Organizations.* New York: Wiley.

Markus, M. L. (2010). On the use of information technology: The history of IT and organizational design in large US enterprises. *Entreprises et Histoire,60*, 17–28.

Markus, M. L., & Loebbecke, C. (2013). Commoditized processes and business community platforms: New opportunities and challenges for digital business strategies. *MIS Quarterly,37* (2), 649–656.

Marquis, C., Lounsbury, M., & Greenwood, R. (2011). Introduction: Community as an institutional order and a type of organizing. In C. Marquis, M. Lounsbury, & R. Greenwood (Eds.), *Communities and organizations, research in the sociology of organizations* (Vol. 33, pp. IX–XXVII). Emerald: Bingley.

Mason, R., McKenney, J. L., & Copeland, D. G. (1997a). An historical method for MIS research: Steps and assumptions. *MIS Quarterly, 21*(3), 307–320.

Mason, R., McKenney, J. L., & Copeland, D. G. (1997b). Developing a historical tradition in MIS research. *MIS Quarterly, 21*(3), 257–278.

Mayer-Schönberger, V., & Cukier, K. (2013). *Big data, a revolution that will transform how we live, work, and think.* New York: Houghton Mifflin Hartcourt Publishing.

McAfee, A. (2009). *Enterprise 2.0, new collaborative tools for your organization's toughest challenges.* MA: Harvard Business Press

McGrath, R. (2013), *The end of competitive advantage: how to keep your strategy moving as fast as your business.* Harvard Business Review Press.

McKenney, J. L., Mason, R., & Copeland, D. G. (1997). The Bank of America: Crest and trough of technological leadership. *MIS Quarterly, 21*(3), 321–353.

McKinsey & Company. (2014). Accélérer la mutation numérique des entreprises: un gisement de croissance et de compétitivité pour la France, Paris.

Méda, D. (2004). *Le travail,* Paris, Presses Universitaires de France, Que sais-je?

Ménard, B. (2010). *L'entreprise numérique, Quelles stratégies pour 2015?* Paris: nuvis.

Morley, C., & Bia-Figuereido, M. (2011). *Génération Y et pratiques de management des projets SI.* Paris: Fondation CIGREF.

Mounier-Kuhn, P. E. (2010). Les clubs d'utilisateurs: entre syndicats de clients, outils de marketing et «logiciel libre» avant la lettre. *Entreprises et Histoire, 60,* 158–10.

Mourmant, G., & Kalika, M. (2012). *Observatory of information systems (is) enterprising individuals.* Paris: CIGREF Foundation.

Mozorov, E. (2011). *The net delusion, the dark side of internet freedom.* New York: PublicAffairs.

Mozorov, E. (2013). *To save everything, click here.* New York: PublicAffairs.

Murata, K. (2010). Lessons from the history of information systems development use in Japan. *Entreprises et Histoire, 60,* 50–61.

Nonaka, I., & Konno, N. (1998). The concept of "BA", building a foundation for knwoledge creation. *California Management Review, 40*(3), 40–54.

OECD. (2014). *Data driven innovation for growth and well-being, interim synthesis report,* Paris, OECD. Aavailable at: http://www.oecd.org/sti/inno/data-driven-innovation-interim-synthesis.pdf

OECD. (2015). *Securing livelihoods for all, foresight for action.* Paris: OECD.

Oestreicher-Singer, G., & Zalmanson, L. (2013). Content or community? A digital business strategy for content providers in the social age. *MIS Quarterly 37*(2), 591–616

Porter, M. E. (1985). *Competitive advantage: Creating and sustaining superior performance.* New York: Free Press.

Poster, M. (1995). *The second media age.* Cambridge, UK: Polity Press.

Ramiller, N., & Swanson, E. B. (2003). Organizing visions for information technology and the information systems executive response. *Journal of Management Information Systems, 20*(1), 13–50.

Ramirez, R. (1999). Value co-production: Intellectual origins and implications for practice and research. *Strategic Management Journal, 20,* 49–65.

Ravidat, N., & Akoka, J. (2006). Evolution du positionnement de la fonction systèmes d'information en France, continuité ou rupture? *Systèmes d'Information et Management, 11*(3), 67.

Ravidat, N., Schmitt, J.-P., & Akoka, J. (2005). *Recherche exploratoire, analyse longitudinale de l'évolution du positionnement de la fonction systèmes d'information de 1992 à 2004.* Cigref

Ricoeur, P. (2004). *Parcours de la reconnaissance, trois études.* Paris: Editions Stock.

Rodhain, F., Fallery, B., Girard, A., & Desq, S. (2010). Une histoire de la recherche en systèmes d'information à travers 30 ans de publications. *Entreprises et Histoire,60*, 78–97.

Rosa, H. (2010). *Accélération. Une critique sociale du temps*. Paris: Editions La Découverte.

Sack, R. D. (1986). *Human territoriality: Its theory and history*. Cambridge: Cambridge University Press.

Salais, R., & Storper, M. (1993). *Les Mondes de production: enquête sur l'identité économique de la France*. Paris: Editions de l'Ecole des hautes études en sciences sociales

Sartori, G. (1998). Homo videns. Televisione e post-pensiero. Firenze, La Nuovo Italia

Seidel, M. D. L., & Stewart, K. J. (2011). An initial description of the C-form. In C. Marquis, M. Lounsbury & Greenwood, R. (Eds.) Communities and organizations, research in the sociology of organizations (Vol. 33, pp. 37–72). Bingley, UK: Emerald.

Simone, R. (2012). *Pris dans la Toile, L'esprit aux temps du web*. Paris: Editions Gallimard.

Stabell, C. B., Fjeldstad, O. D. (1998). "Configuring value for competitive advantage, on chains, shops and networks". *Strategic Management Journal, 19*, 413–437.

Swanson, E. B., & Ramiller, N. C. (1997). The organizing vision in information systems innovation. *Organization Science,8*(5), 458–474.

The World Bank. (2006). *Where is the wealth of nations*. Washington DC.

Thompson, J. D. (1967). *Organizations in action*. New York: McGraw-Hill.

Ting, Peng L. (2012). *How information technologies affect the knowledge ecology and their adoption of open innovation: A multinational study*. Paris: CIGREF Foundation.

Tönnies, F. (1977). *Communauté et société: catégories fondamentales de la sociologie pure. Communauté et société*. Paris: Retz-Centre d'Études et de Promotion de la Lecture. Available at: http://classiques.uqac.ca/classiques/tonnies_ferdinand/tonnies.html

Touraine, A. (1973). *Production de la société*. Paris: Editions Le Seuil.

Tran, S. (Ed.). (2011). *L'impact du Web 2.0 sur les organisations*. Springer, Coll. Espaces numériques.

Tuomi, I. (2002). *Networks of innovation*. New York: Oxford University Press.

Tushman, M., Lakhani, K., & Lifshitz-Assaf, H. (2012). Open innovation and organization design. *Organization Design, JOD,1*(1), 24–27.

Verley, P. (1997). *La révolution industrielle*. Paris: Gallimard.

Williamson, O. E. (1975). *Markets and hierarchies, analysis and ainttrust implications, a study in the economics of internal organization*. New York: Free Press.

Williamson, O. E. (1981). *The economics of organizations: The transaction cost approach*. New York: Free Press.

Woetzel, et al. (2014). *China's digital transformation: The Internet impact on productivity and growth*. New York: McKinsey & Company.

Womack, J. P., Jones, D. T., Roos, D. (1991). *The machine that changed the world*. New York: Harper Perennial Edition.

Woodard, C. J., Ramasubbu, N., Tschang, F. T. & Sambamurthy, V. (2013). Design capital and design moves: The logic of digital business strategy. *MIS Quarterly, 37*(2), 537–564.

Yates, J. (2005). *Structuring the information age: Life insurance and technology in the twentieth century*. Baltimore: Johns Hopkins University Press.

Zittrain, J. (2006). The Generative Internet, *Harvard Law Review, 119*, 1074–2040.

List of Final Reports for the ISD Program

Abbot, P., Zheng, Y. (2012). *Globally distributed innovation and co-creation of value: Cases of UK-China collaborations*. Paris: CIGREF Foundation.

Adams, A., Muarata, K. (2013). *An East Asian perspective on the developing ethical and social values of digital object usage*. Paris: CIGREF Foundation.

Aldebert, B., Bertrand, D., Monnoyer, M.-C., Seck A.-M. (2014). *Les systèmes d'information inter-organisationnels et la création de valeur : résultats d'une étude menée auprès de deux grandes entreprises françaises Eurocopter et Cegelec-Vinci et leurs fournisseurs*. Paris: Fondation CIGREF.

Benbya, H. (2013). *Exploring the design and effects of internal knowledge markets*. Paris: CIGREF Foundation.

Casilli, A., Tubaro, P. (2012). *Testing the "end of privacy" hypothesis in computer-mediated communication: An agent-based modelling approach*. Paris: CIGREF Foundation.

Cho, N. (2011). *The use of smart mobile equipment for the innovation in organizational coordination*. Paris: CIGREF Foundation.

Collins, P., Majchrzak, A. (2014). *Accelerating innovation through collaborative co-creation in 2020 R&D organizations*. Paris: CIGREF Foundation.

De Vaujany, F.-X., Carton, S., Dominguez, C. (2011). *Pratiques émergentes et systèmes d'information: une perspective multi-niveaux*. Paris: Fondation CIGREF.

De Vita, R., Conaldi, G. (2014). *Re-defining the space for companies-communities interaction: How can firms leverage the innovative potential of open source software production model?* Paris: CIGREF Foundation.

De Vreede, G.-J., Boughzala, I., Nguyen, C., de Vreede, T., Oh, O. (2014). *Towards a maturity model for the assessment of ideation processes in crowdsourcing projects*. Paris: CIGREF Foundation.

Delmond, M.-H., Keravel, A., Coelho, F., Mahl, R. (2011). *Business models, coproduction de valeur et systèmes d'information*. Paris: Fondation CIGREF.

Dupouët, O., Bouzdine-Chameeva, T., & Lakshman, C. (2011). *Innovation from information systems: An ambidexterity approach*. Paris: CIGREF Foundation.

Dutta, A., Menon, N. M. (2013). *IT configurations and organizational performance*. Paris: CIGREF Foundation.

El Sawy, O. A., Pereira, F. (2011). *Towards a unified framework for business modelling in the evolving digital space*. Paris: CIGREF Foundation.

El Sawy, O. A., Pereira, F., Ploof, R. (2014). *Rethinking enterprise 2020 for networked abundance in the value-shifting age of the internet of things*. Paris: CIGREF Foundation.

Elie Dit Cosaque, C., Pallud, J. (2014). *Changement et adaptation individuelle aux innovations organisationnelles et aux technologies émergentes: Adoption, meilleures pratiques et performance*. Paris: Fondation CIGREF.

Ge, C. (2013). *Intellectual property law and freedom: Between the national and the international*. Paris: CIGREF Foundation.

© Springer International Publishing Switzerland 2016
A. Bounfour, *Digital Futures, Digital Transformation*,
Progress in IS, DOI 10.1007/978-3-319-23279-9

Greenan, N., Gillet, I., Le Gall, R. (2014). *Centre d'appel 2020*. Centre d'Etudes de l'Emploi. Paris: Fondation CIGREF.

Jacobs, K. (2013). *Corporate standardisation management in the ICt sector*. Paris: CIGREF Foundation.

Majchrzak, A., Pereira, F., Fife, E., Johnson, J., Qingfei, M. (2012). *An action program for guiding and assessing the use of social media tools in the enterprise enhancing collaborative innovation, collective problem solving and knowledge sharing*. Paris: CIGREF Foundation.

Morley, C., Bia-Figuereido, M. (2011). *Génération Y et pratiques de management des projets SI*. Paris: Fondation CIGREF.

Mourmant, G., Kalika, M. (2012). *Observatory of information systems (IS) enterprising individuals*. Paris: CIGREF Foundation.

Oiry, E., Ologeanu-Taddei, R., Pascal, A., Tchobanian, R., Fallery, B. (2011). *Développer les usages des logiciels collaboratifs, le rôle des SI, des RH et des managers*. Paris: Fondation CIGREF.

Reimers, K., Guo, X., Li, M., Xie, B. (2013). *Innovating in a learning community*. Paris: CIGREF Foundation.

Soulier, E., Silva, F., Maged, H. (2011). *Définir et évaluer une nouvelle méthodologie s'appuyant sur des technologies innovantes pour étudier des pratiques émergentes dans les activités professionnelles*. Paris: Fondation CIGREF.

Ting Peng L. (2012). *How information technologies affect the knowledge ecology and their adoption of open innovation: A multinational study*. Paris: CIGREF Foundation.

Tran, S. (sous la dir.). (2011). *L'impact du Web 2.0 sur les organisations*. Springer: Coll. Espaces numériques.

Wakunuma, K., Masclet, L., Bernd, S., Goujon, P., Sara, W. (2013). *Identification and governance of emerging ethical issues in information systems*. Paris: CIGREF Foundation.

Yoo, Y., Kulathinal, R., Wattal, S. (2014). *Designing 21st century organizations for generativity: An organizational genetics approach*. Paris: CIGREF Foundation.